Mohr Circles, Stress Paths and Geotechnics

Second Edition

Mohr Circles, Stress Paths and Geotechnics

Second Edition

R H G Parry

CRC Press
Taylor & Francis Group
Boca Raton London New York

CRC Press is an imprint of the
Taylor & Francis Group, an **informa** business

CRC Press
Taylor & Francis Group
6000 Broken Sound Parkway NW, Suite 300
Boca Raton, FL 33487-2742

First issued in paperback 2019

ISBN-13: 978-0-415-27297-1 (hbk)
ISBN-13: 978-0-367-87125-3 (pbk)

Visit the Taylor & Francis Web site at
http://www.taylorandfrancis.com

and the CRC Press Web site at
http://www.crcpress.com

British Library Cataloguing in Publication Data
A catalogue record for this book is available from the British Library

Library of Congress Cataloging in Publication Data
A catalog record has been requested

Contents

Worked examples

Preface

On turning the pages of the many textbooks which already exist on soil mechanics and rock mechanics, the important roles of Mohr circles and stress paths in geotechnics becomes readily apparent. They are used for representing and interpreting data, for the analysis of geotechnical problems and for predicting soil and rock behaviour. In the present book, Mohr circles and stress paths are explained in detail – including the link between Mohr stress circles and stress paths – and soil and rock strength and deformation behaviour are viewed from the vantage points of these graphical techniques. Their various applications are drawn together in this volume to provide a unifying link to diverse aspects of soil and rock behaviour. The reader can judge if the book succeeds in this.

Past and present members of the Cambridge Soil Mechanics Group will see much in the book which is familiar to them, as I have drawn, where appropriate, on the accumulated Cambridge corpus of geotechnical material in the form of reports, handouts and examples. Thus, a number of people have influenced the contents of the book, and I must take this opportunity to express my gratitude to them. I am particularly grateful to Ian Johnston and Malcolm Bolton for looking through the first drafts and coming up with many useful suggestions. Similarly my thanks must go to the anonymous reviewers to whom the publishers sent the first draft for comments, and who also came up with some very useful suggestions. The shortcomings of the book are entirely of my own making.

Inevitably, there has been much typing and retyping, and I owe a special debt of gratitude to Stephanie Saunders, Reveria Wells and Amy Cobb for their contributions in producing the original manuscript. I am also especially indebted to Ulrich Smoltczyk, who provided me with a copy of Mohr's 1882 paper and an excellent photograph of Mohr, and to Stille Olthoff for translating the paper. Markus Caprez kindly assisted me in my efforts to locate a copy of Culmann's early work.

Permissions to reproduce diagrams in this text have been kindly granted by The American Society of Civil Engineers, The American Society of Mechanical Engineers, Dr B. H. Brady, Dr E. W. Brooker, Dr F. A. Donath, Elsevier Science Publishers, Dr R. E. Gibson, McGraw-Hill Book Co., Sociedad Española de Mecanica del Suelo y Cimentaciones, and Thomas Telford Services Limited.

R H G Parry
Cambridge

Historical note: Karl Culmann (1821–1881) and Christian Otto Mohr (1835–1918)

Although the stress circle is invariably attributed to Mohr, it was in fact Culmann who first conceived this graphical means of representing stress. Mohr's contribution lay in making an extended study of its usage for both two-dimensional and three-dimensional stresses, and in developing a strength criterion based on the stress circle at a time when most engineers accepted Saint-Venant's maximum strain theory as a valid failure criterion. Anyone wishing to pursue the relative contributions of Culmann and Mohr is recommended to read the excellent accounts in *History of Strength of Materials* by Timoshenko (McGraw-Hill, 1953).

Born in Bergzabern, Rheinpfalz, in 1821, Karl Culmann graduated from the Karlsruhe Polytechnikum in 1841 and immediately started work at Hof on the Bavarian railroads. In 1849 the Railways Commission sent him to England and the United States for a period of two years to study bridge construction in those countries. The excellent engineering education which he had received enabled him to view, from a theoretical standpoint, the work of his English and American counterparts, whose expertise was based largely on experience. The outcome was a report by Culmann published in 1852 which strongly influenced the theory of structures and bridge engineering in Germany. His appointment as Professor of Theory of Structures at the Zurich Polytechnikum in 1855 gave him the opportunity to develop and teach his ideas on the use of graphical methods of analysis for engineering structures, culminating in his book *Die Graphische Statik*, published by Verlag von Meyer and Zeller in 1866. The many areas of graphical statics dealt with in the book include the application of the polygon of forces and the funicular polygon, construction of the bending moment diagram, the graphical solution for continuous beams (later simplified by Mohr) and the use of the method of Sections for analysing trusses. He concluded this book with Sections on calculating the pressures on retaining walls and tunnels.

Culmann introduced his stress circle in considering longitudinal and vertical stresses in horizontal beams during bending. Isolating a small element of the beam and using rectangular coordinates, he drew a circle with its centre on the (horizontal) zero shear stress axis, passing through the two stress points represented by the normal and conjugate shear stresses on the vertical and horizontal faces of the element. He took the normal stress on the horizontal faces to be zero. In making

Figure 1 Karl Culmann.

Figure 2 Otto Mohr.

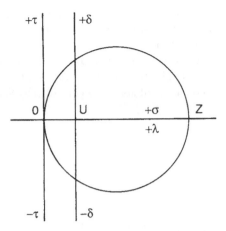

Figure 3 A figure from Mohr's 1882 paper showing his use of a single circle to illustrate both stress and strain circles for uniaxial tension.

this construction Culmann established a point on the circle, now known as the pole point, and showed that the stresses on a plane at any specified inclination could be found by a line through this point drawn parallel to the plane. Such a line met the circle again at the required stress point. Extensive use is made of the pole point in the present text. Culmann went on to plot trajectories of principal stresses for a beam, obtained directly from the stress circles.

Christian Otto Mohr was born in 1835 in Wesselburen, on the inhospitable North Sea coast of Schleswig-Holstein. After graduating from the Hannover Polytechnical Institute he first worked, like Culmann, as a railway engineer before taking up, at the age of 32, the post of Professor of Engineering Mechanics at the Stuttgart Polytechnikum. In 1873 he moved to the Dresden Polytechnikum, where he continued to pursue his interests in both strength of materials and the theory of structures. Pioneering contributions which he made to the theory of structures included the use of influence lines to calculate the deflections of continuous beams, a graphical solution to the three-moments equations, and the concept of virtual work to calculate displacements at truss joints. His work on the stress circle included both two-dimensional and three-dimensional applications and, in addition, he formulated the trigonometrical expressions for an elastic material, relating stresses and strains, as well as the expression relating direct and shear strain moduli. As with stress, he showed that shear strains and direct strains could be represented graphically by circles in a rectangular coordinate system.

Believing, as Coulomb had done a hundred years before, that shear stresses caused failure in engineering materials, Mohr proposed a failure criterion based on envelopes tangential to stress circles at fracture in tension and compression. He then assumed that any stress conditions represented by a circle touching these envelopes would initiate failure. This failure criterion was found to give better agreement with experiment than the maximum strain theory of Saint-Venant, which was widely accepted at that time.

Mohr first published his work on stress and strain circles in 1882 in *Civilingenieur* and it was repeated in *Abhandlungen aus dem Gebiete der Technischen Mechanik* (2nd edn), a collection of his works published by Wilhelm Ernst & Sohn, Berlin, 1914.

Chapter 1

Stresses, strains and Mohr circles

1.1 The concept of stress

The concept of stress, defined as force per unit area, was introduced into the theory of elasticity by Cauchy in about 1822. It has become universally used as an expedient in engineering design and analysis, despite the fact that it cannot be measured directly and gives no indication of how forces are transmitted through a stressed material. Clearly the manner of transfer in a solid crystalline material, such as a metal or hard rock, is different from the point-to-point contacts in a particulate material, such as a soil. Nevertheless, in both cases it is convenient to visualize an imaginary plane within the material and calculate the stress across it by simply dividing the force across the plane by the total area of the plane.

1.2 Simple axial stress

A simple illustration of stress is given by considering a cylindrical test specimen, with uniform Section of radius r, subjected to an axial compressive force F as shown in Figure 1.1(a). Assuming the force acts uniformly across the Section of the specimen, the stress σ_{n0} on a plane PQ perpendicular to the direction of the force, as shown in Figure 1.1(a), is given by

$$\sigma_{n0} = \frac{F}{A} \tag{1.1}$$

where A is the cross-sectional area of the specimen. As this is the only stress acting across the plane, and it is perpendicular to the plane, σ_{n0} is a principal stress.

Consider now a plane such as PR in Figure 1.1(b), inclined at an angle θ to the radial planes on which σ_{n0} acts. The force F has components N acting normal (perpendicular) to the plane and T acting along the plane, in the direction of maximum inclination θ. Thus

$$N = F \cos \theta \tag{1.2a}$$

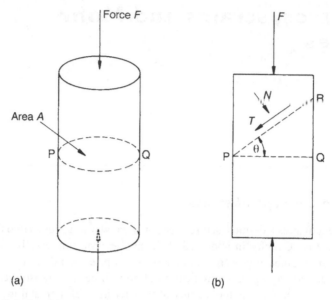

Figure 1.1 Cylindrical test specimen subjected to axial force F.

$$T = F \sin \theta \qquad (1.2b)$$

As the inclined plane is an ellipse with area $A/\cos\theta$, the direct stress $\sigma_{n\theta}$ normal to the plane and shear stress τ_θ along the plane, in the direction of maximum inclination, are given by:

$$\sigma_{n\theta} = \frac{N\cos\theta}{A} = \frac{F}{A}\cos^2\theta \qquad (1.3a)$$

$$\tau_\theta = \frac{T\cos\theta}{A} = \frac{F}{2A}\sin 2\theta \qquad (1.3b)$$

It is obvious by inspection that the maximum normal stress, equal to F/A, acts on radial planes. The magnitude and direction of the maximum value of τ_θ can be found by differentiating equation 1.3b:

$$\frac{d\tau_\theta}{d\theta} = \frac{F}{A}\cos 2\theta$$

The maximum value of τ_θ is found by putting $d\tau_\theta/d\theta = 0$, thus:

$$\cos 2\theta = 0$$
$$\theta = 45°\left(\text{or } 135°\right) \qquad (1.4)$$
$$\tau_{\theta max} = \frac{F}{2A}$$

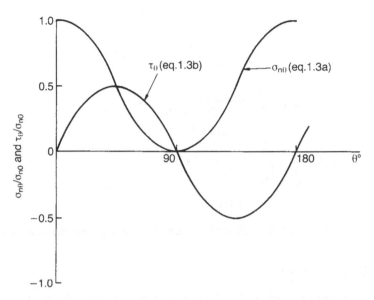

Figure 1.2 Variation of normal stress $\sigma_{n\theta}$ and shear stress τ_{θ} with angle of plane θ in cylindrical test specimen.

The variations of $\sigma_{n\theta}$ and τ_{θ} with θ, given by equations 1.3a and 1.3b, are shown in Figure 1.2. It can be seen that $\tau_{\theta max}$ occurs on a plane with $\theta = 45°$ and $\sigma_{n\theta max}$ on a plane with $\theta = 0°$.

EXAMPLE 1.1 AXIAL PRINCIPAL STRESSES

A cylindrical specimen of rock, 50 mm in diameter and 100 mm long is subjected to an axial compressive force of 5 kN. Find:

1. the normal stress $\sigma_{n\theta}$ and shear stress τ_{θ} on a plane inclined at 30° to the radial direction;

2. the maximum value of shear stress;
3. the inclination of planes on which the shear stress τ_θ is equal to one-half $\tau_{\theta\max}$.

Solution

1. Area $A = \pi r^2 = \pi \times 0.025^2 \text{ m}^2 = 1.96 \times 10^{-3} \text{ m}^2$

$$\text{Equation 1.3a: } \sigma_{n\theta} = \frac{5}{1.96 \times 10^{-3}} \times \cos^2 30° = 1913 \text{ kPa}$$

$$\text{Equation 1.3b: } \tau_\theta = \frac{5}{2 \times 1.96 \times 10^{-3}} \times \sin 60° = 1105 \text{ kPa}$$

2. Equation 1.4: $\tau_{\theta\max} = \dfrac{F}{2A} = \dfrac{5}{2 \times 1.96 \times 10^{-3}} \text{ kPa}$

$$= 1275 \text{ kPa}$$

3. Equation 1.3b: $\frac{1}{2}\tau_{\theta\max} = \tau_{\theta\max} \sin 2\theta$

$$\therefore \ \sin 2\theta = \tfrac{1}{2}$$

$$\therefore \theta = 15° \text{ or } 75°$$

1.3 Biaxial stress

Although in most stressed bodies the stresses acting at any point are fully three-dimensional, it is useful for the sake of clarity to consider stresses in two dimensions only before considering the full three-dimensional stress state.

1.3.1 Simple biaxial stress system

A simple biaxial stress system is shown in Figure 1.3(a), which represents a rectangular plate of unit thickness with stresses σ_1, σ_2 acting normally to the squared edges of the plate. As the shear stresses along the edges are assumed to be zero, σ_1 and σ_2 are principal stresses.

A small square element of the plate is shown in the two-dimensional diagram in Figure 1.3(b). The stresses $\sigma_{n\theta}$, τ_θ acting on a plane inclined at an angle θ to the direction of the plane on which σ_1 acts can be found by considering the forces acting on the triangular element in Figure 1.3(c).

If length $CD = l$, then for a plate of unit thickness:

$$F_1 = \sigma_1 l \tag{1.5a}$$

$$N_1 = \sigma_1 l \cos \theta \tag{1.5b}$$

$$T_1 = \sigma_1 l \sin \theta \tag{1.5c}$$

$$F_2 = \sigma_2 l \tan \theta \tag{1.5d}$$

$$N_2 = \sigma_2 l \tan \theta \sin \theta \tag{1.5e}$$

$$T_2 = \sigma_2 l \tan \theta \cos \theta \tag{1.5f}$$

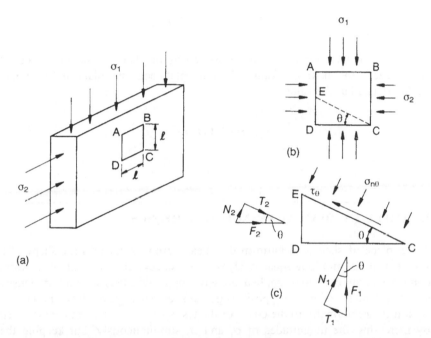

Figure 1.3 Biaxial stress system in a rectangular plate: (a) boundary stresses; (b) stresses on element ABCD; (c) determination of stresses $\sigma_{n\theta}$, τ_θ on plane inclined at angle θ.

Resolving forces in the direction of action of $\sigma_{n\theta}$:

$$\sigma_{n\theta} l \sec\theta = N_1 + N_2 \tag{1.6}$$

Substituting equations 1.5b, 1.5e into equation 1.6:

$$\sigma_{n\theta} = \sigma_1 \cos^2 \theta + \sigma_2 \sin^2 \theta \tag{1.7}$$

Resolving forces in the direction of τ_θ:

$$\tau_\theta / \sec\theta = T_1 + T_2 \tag{1.8}$$

Substituting equations 1.5c, 1.5f into equation 1.8:

$$\tau_\theta = \tfrac{1}{2}\left(\sigma_1 - \sigma_2\right)\sin 2\theta \tag{1.9}$$

Comparing equation 1.9 with equation 1.3b it is seen that the maximum value of τ_θ acts on a plane with $\theta = 45°$, for which

$$\tau_{\theta max} = \tfrac{1}{2}\left(\sigma_1 - \sigma_2\right) \tag{1.10}$$

This is not the maximum value of shear stress in the plate. As the third principal stress is zero, the maximum value of τ in the plate acts on a plane at 45° to both σ_1 and σ_2 and has the value

$$\tau_{max} = \frac{\sigma_1}{2} \text{ if } \sigma_1 > \sigma_2$$

EXAMPLE 1.2 BIAXIAL PRINCIPAL STRESSES

A flat piece of slate with uniform thickness 20 mm is cut into the shape of a square with 100 mm long squared edges. A test is devised which allows uniform compressive stress σ_1 to be applied along two opposite edges and uniform tensile stress σ_2 along the other two opposite edges, as shown in Figure 1.4(a). The stresses σ_1 and σ_2 act normally to the edges of the test specimen. The test is performed by increasing the magnitudes of σ_1 and σ_2 simultaneously, but keeping the magnitude of σ_1 always four times the magnitude of σ_2. If failure of the slate occurs when the shear stress on any plane exceeds 1 MPa, what would be the values of σ_1 and σ_2 at the moment of failure?

Would the values of σ_1 and σ_2 at failure be changed if:

1. the rock had a tensile strength of 0.5 MPa?
2. a planar weakness running through the test specimen as shown in Figure 1.4(b), inclined at 60° to the direction of σ_2, would rupture if the shear stress on it exceeds 0.8 MPa?

Solution

The maximum shear stress $\tau_{\theta max}$ occurs on a plane with $\theta = 45°$.

Equation 1.10: $$\tau_{\theta max} = \tfrac{1}{2}(\sigma_1 - \sigma_2)$$

As $\tau_{\theta max} = 1$ MPa and $\sigma_1 = -4\sigma_2$, then at failure:

$$\sigma_1 = 1.6\,\text{MPa} \qquad \sigma_2 = -0.4\,\text{MPa}$$

1. As σ_2 at failure is less in magnitude than 0.5 MPa, the tensile strength of the slate does not influence failure.
2. As the planar weakness acts at 30° to the direction of σ_1, the normal stress on this weakness $\sigma_{n\theta}$ acts in a direction of 60° to σ_1. Thus the stress τ_θ on the plane of weakness is found by putting $\theta = 60°$.

Equation 1.9:

$$\tau_\theta = \tfrac{1}{2}(\sigma_1 - \sigma_2)\sin 2\theta$$

If $\theta = 60°$,

$$\tau_\theta = 0.866\,\text{MPa}$$

As rupture occurs on the plane of weakness when $\tau_\theta = 0.8$ MPa, the weakness would influence failure. The stresses at rupture are found by putting $\tau = 0.8$ MPa and $\sigma_1 = -4\sigma_2$ into equation 1.9, giving

$$\sigma_1 = 1.48\,\text{MPa} \qquad \sigma_2 = -0.37\,\text{MPa}$$

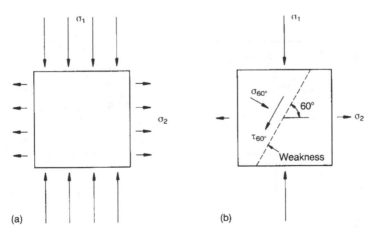

Figure 1.4 **Example 1.2.**

1.3.2 Generalized biaxial stress system

In Section 1.3.1 the special case is considered of an element ABCD with principal stresses only acting along its edges, but it has been seen that shear stresses are generated along all other planes which are not parallel to the normally loaded edges of the plate. Thus, an element with edges oriented in directions z, y different from the principal stress directions will have both normal and direct stresses acting along its edges, as shown in Figure 1.5(a).

It is appropriate here to consider the conventions of stress representation usually adopted in geotechnics. Referring to Figure 1.5(a) we have the following.

1. Shear stress τ_{zy} acts tangentially along the edge or face normal to the direction z and in the direction y. The converse applies for τ_{yz}.
2. Compressive normal stresses are positive and tensile normal stresses are negative.
3. Anticlockwise shear stresses are positive and clockwise shear stresses are negative. Thus, in Figure 1.5(a):

$$\tau_{zy} \text{ is +ve}$$

$$\tau_{yz} \text{ is -ve}$$

If moments are taken about a point such as M in Figure 1.5(a), static equilibrium of the element JKLM can be maintained if the conjugate shear stresses τ_{zy} and τ_{yz} are equal in magnitude, i.e.

$$\tau_{zy} = -\tau_{yz} \tag{1.11}$$

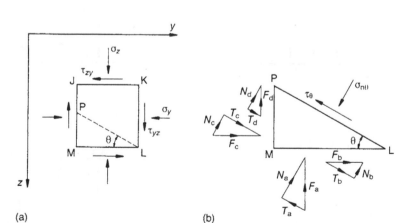

(a) (b)

Figure 1.5 Generalized biaxial stress system: (a) stresses on element JKLM; (b) determination of stresses $\sigma_{n\theta}$, τ_θ on plane inclined at angle θ.

This is known as the principle of complementary shear.

The stresses on a plane inclined at angle θ to the direction of the plane on which σ_1 acts can be obtained by considering the equilibrium of the triangular portion LMP of element JKLM shown in Figure 1.5(b). The forces acting on edges LM and MP can be determined from the force diagrams:

$$
\begin{array}{lll}
F_a = \sigma_z l & N_a = F_a \cos \theta & T_a = F_a \sin \theta \\
F_b = \tau_{zy} l & N_b = F_b \sin \theta & T_b = F_b \cos \theta \\
F_c = \sigma_y l \tan \theta & N_c = F_c \sin \theta & T_c = F_c \cos \theta \\
F_d = -\tau_{yz} l \tan \theta & N_d = F_d \cos \theta & T_d = F_d \sin \theta
\end{array}
$$

But, noting that $-\tau_{yz} = \tau_{zy}$

$$
F_d = -\tau_{yz} l \tan\theta = \tau_{zy} l \tan\theta
$$

Resolving forces in the direction of $\sigma_{n\theta}$:

$$
\sigma_{n\theta} l \sec \theta = N_a + N_b + N_c + N_d
$$

which becomes, on substitution of the force diagram values:

$$
\sigma_{n\theta} = \sigma_z \cos^2 \theta + \sigma_y \sin^2 \theta + \tau_{zy} \sin 2\theta \tag{1.12}
$$

Resolving forces in the direction of τ_θ:

$$
\tau_\theta l \sec \theta = -T_a + T_b + T_c + T_d
$$

which becomes, on substitution of the force diagram values:

$$
\tau_\theta = \tfrac{1}{2}\left(\sigma_y - \sigma_z\right)\sin 2\theta + \tau_{zy} \cos 2\theta \tag{1.13}
$$

(a) Planes on which $\tau_\theta = 0$

The directions of planes on which $\tau_\theta = 0$ can be found by putting $\tau_\theta = 0$ in equation 1.13, from which

$$
\tan 2\theta = \frac{2\tau_{zy}}{\sigma_z - \sigma_y} \tag{1.14}
$$

Equation 1.14 gives two sets of orthogonal planes. As the shear stress is zero on these planes, these are the planes on which the principal stresses act.

It is possible to evaluate the principal stresses on these planes by substituting equation 1.14 into equation 1.12, noting that equation 1.14 gives

$$\sin 2\theta = \frac{2\tau_{zy}}{\left[\left(\sigma_z - \sigma_y\right)^2 + 4\tau_{zy}^2\right]^{1/2}}$$

$$\cos 2\theta = \frac{\sigma_z - \sigma_y}{\left[\left(\sigma_z - \sigma_y\right)^2 + 4\tau_{zy}^2\right]^{1/2}}$$

and using the trigonometrical relationships

$$\cos^2 \theta = \frac{1 + \cos 2\theta}{2}$$

$$\sin^2 \theta = \frac{1 - \cos 2\theta}{2}$$

Substitution of these relationships into equation 1.12 gives

$$\sigma_{n\theta} = \tfrac{1}{2}\left(\sigma_z + \sigma_y\right) \pm \tfrac{1}{2}\left[\left(\sigma_z - \sigma_y\right)^2 + 4\tau_{zy}^2\right]^{1/2} \tag{1.15}$$

As the expression under the root sign yields both positive and negative values, two values of $\sigma_{n\theta}$ are obtained, the larger of which is the major and the smaller the minor of the two principal stresses.

(b) Planes on which maximum τ_θ acts

The directions of planes on which the maximum values of τ_θ act can be found by differentiating equation 1.13 with respect to θ and equating to zero:

$$\tan 2\theta = \frac{\sigma_y - \sigma_z}{2\tau_{zy}} \tag{1.16}$$

It is possible to evaluate the maximum shear stress $\tau_{\theta\text{max}}$ by substituting equation 1.16 into equation 1.13:

$$\tau_{\theta\text{max}} = \tfrac{1}{2}\left[\left(\sigma_y - \sigma_z\right)^2 + 4\tau_{zy}^2\right]^{1/2} \tag{1.17}$$

In the mathematical sense $\tau_{\theta\max}$ is the positive root of equation 1.17, while the negative root gives the minimum value of τ_θ, which has the same magnitude as, but is opposite in sense to, $\tau_{\theta\max}$. In the physical sense the minimum value of τ_θ is zero.

1.4 Mohr stress circle

A graphical means of representing the foregoing stress relationships was discovered by Culmann (1866) and developed in detail by Mohr (1882), after whom the graphical method is now named.

By using the relationships

$$\cos^2\theta = \frac{1+\cos 2\theta}{2}$$

$$\sin^2\theta = \frac{1-\cos 2\theta}{2}$$

it is possible to rewrite equation 1.12 in the form

$$\sigma_{n\theta} - \tfrac{1}{2}(\sigma_z + \sigma_y) = \tfrac{1}{2}(\sigma_z - \sigma_y)\cos 2\theta + \tau_{zy}\sin 2\theta \qquad (1.18)$$

If equations 1.13 and 1.18 are squared and added, after some manipulation (remembering that $\sin^2 2\theta + \cos^2 2\theta = 1$) the following expression results:

$$\left[\sigma_{n\theta} - \tfrac{1}{2}(\sigma_z + \sigma_y)\right]^2 + \tau_\theta^2 = \left[\tfrac{1}{2}(\sigma_z - \sigma_y)\right]^2 + \tau_{zy}^2 \qquad (1.19)$$

Putting

$$s = \tfrac{1}{2}(\sigma_z + \sigma_y)$$

$$r^2 = \left[\tfrac{1}{2}(\sigma_z - \sigma_y)\right]^2 + \tau_{zy}^2$$

equation 1.19 becomes

$$\left(\sigma_{n\theta} - s\right)^2 + \tau_\theta^2 = r^2$$

which is the equation of a circle with radius r and with a centre, on a τ–σ plot, at

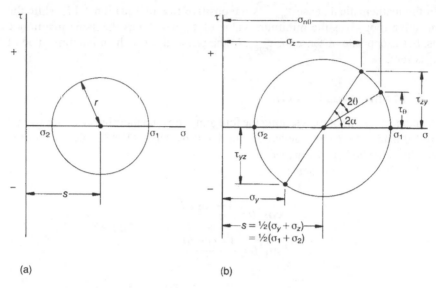

Figure 1.6 Mohr stress circle: (a) geometry; (b) stress representation.

the point $\sigma = s$, $\tau = 0$, as shown in Figure 1.6(a). This is the Mohr circle of stress. The complete state of two-dimensional stress is represented by points on this circle. The principal stresses σ_1, σ_2 are given by the points where the circle crosses the $\tau = 0$ axis.

In Figure 1.6(b) the boundary stresses σ_z, σ_y, τ_{zy}, τ_{yz}, on the element shown in Figure 1.5 are plotted in τ–σ space and a circle drawn through them. The Mohr circle represents completely the two-dimensional stresses acting within the element, and σ_1, σ_2 are the major and minor principal stresses respectively. The stresses $\sigma_{n\theta}$, τ_θ acting on a plane at an angle θ clockwise to the plane on which σ_z acts, as shown in Figure 1.5, can be found by travelling clockwise around the circle from stress point σ_z, τ_{zy} a distance subtending an angle 2θ at the centre of the circle. Thus, the major principal stress σ_1 acts on a plane inclined at an angle α to the plane on which σ_z acts.

The stresses $\sigma_{n\theta}$, τ_θ can be evaluated from the known boundary stresses σ_z, σ_y, τ_{zy} by observing from Figure 1.6(b) that

$$\sigma_{n\theta} = \tfrac{1}{2}\left(\sigma_z + \sigma_y\right) + r\cos\left(2\alpha - 2\theta\right) \tag{1.20}$$

Substituting

$$r = \frac{\tau_{zy}}{\sin 2\alpha} \tag{1.21a}$$

and

$$\tan 2\alpha = \frac{2\tau_{zy}}{\sigma_z - \sigma_y} \tag{1.21b}$$

obtained from Figure 1.6(b) into equation 1.20 leads to the expression

$$\sigma_{n\theta} = \tfrac{1}{2}\left(\sigma_z + \sigma_y\right) + \tfrac{1}{2}\left(\sigma_z - \sigma_y\right)\cos 2\theta + \tau_{zy}\sin 2\theta$$

This expression is identical to equation 1.18 as required.
  Similarly, from Figure 1.6(b):

$$\tau_\theta = r\sin\left(2\alpha - 2\theta\right) \tag{1.22}$$

Substituting equations 1.21a and 1.21b into equation 1.22 leads to the expression

$$\tau_\theta = \tfrac{1}{2}\left(\sigma_y - \sigma_z\right)\sin 2\theta + \tau_{zy}\cos 2\theta$$

This expression is identical to equation 1.13 as required.
  It can also be seen in Figure 1.6(b) that

$$s = \tfrac{1}{2}\left(\sigma_y + \sigma_z\right) = \tfrac{1}{2}\left(\sigma_1 + \sigma_2\right)$$

1.5 Mohr circles for simple two-dimensional stress systems

Examples of Mohr circles for simple two-dimensional stress systems are shown in Figure 1.7. As stresses in only two dimensions are considered, the diagrams are incomplete. The complete three-dimensional stress diagrams are discussed in Section 1.6, but it is instructive to consider firstly stresses in two dimensions only.

1. **Biaxial compression.** The biaxial stresses are represented by a circle which plots in positive σ space, passing through stress points σ_1, σ_2, on the $\tau = 0$ axis. The centre of the circle is located on the $\tau = 0$ axis at stress point $\tfrac{1}{2}(\sigma_1 + \sigma_2)$. The radius of the circle has the magnitude $\tfrac{1}{2}(\sigma_1 - \sigma_2)$, which is equal to τ_{max}.
2. **Biaxial compression/tension.** In this case the stress circle extends into both positive and negative σ space. The centre of the circle is located on the $\tau = 0$ axis at stress point $\tfrac{1}{2}(\sigma_1 + \sigma_2)$ and has radius $\tfrac{1}{2}(\sigma_1 - \sigma_2)$. This is also the

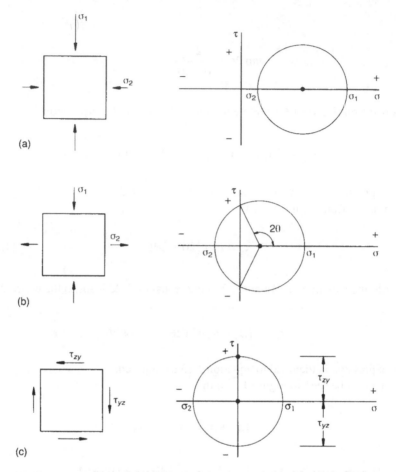

Figure 1.7 Simple biaxial stress systems: (a) compression; (b) tension/compression; (c) pure shear.

maximum value of shear stress, which occurs in a direction at 45° to the σ_1 direction. The normal stress is zero in directions $\pm\,\theta$ to the direction of σ_1, where

$$\cos 2\theta = -\frac{\sigma_1+\sigma_2}{\sigma_1-\sigma_2} \tag{1.23}$$

3. **Biaxial pure shear**. In this case the circle has a radius equal to τ_{zy}, which is equal in magnitude to τ_{yz}, but opposite in sign. The centre of the circle is at $\sigma=0$, $\tau=0$. The principal stresses σ_1, σ_2 are equal in magnitude, but opposite in sign, and are equal in magnitude to τ_{zy}. The directions of σ_1, σ_2 are at 45° to the directions of τ_{zy}, τ_{yz}.

EXAMPLE 1.3 MOHR CIRCLE FOR TWO-DIMENSIONAL STRESSES

Draw the Mohr circle for the stresses at failure for the test specimen in Example 1.2 assuming it has a uniform shear strength of 1 MPa. Determine the directions of zero normal stress and the magnitude of the shear stress acting along the planar weaknesses at 30° to the direction of σ_1.

Solution

At failure

$$\sigma_1 = 1.6 \, \text{MPa}$$
$$\sigma_2 = -0.4 \, \text{MPa}$$

These values give the plot shown in Figure 1.8.

Zero normal stress acts in the directions $\pm\theta$ to the direction of σ_1, where 2θ is given by equation 1.23. Thus

$$\cos 2\theta = -\frac{1.6 + (-0.4)}{1.6 - (-0.4)}$$

$$\therefore \quad \cos 2\theta = -0.6$$

$$\therefore \quad 2\theta = 126.9°$$

$$\therefore \quad \theta = 63.5°$$

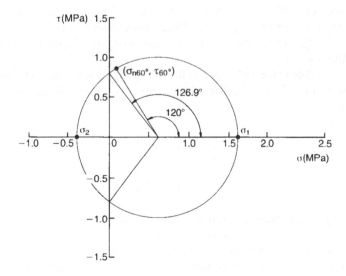

Figure 1.8 Example 1.3.

The angle 2θ can also be found by direct measurement from Figure 1.8.

As the planar weakness acts at $30°$ to the direction of σ_1, the normal stress $\sigma_{n\theta}$ on this plane acts in a direction of $60°$ to the direction of σ_1. Thus, the stresses $\sigma_{n60°}$, $\tau_{60°}$ acting on this plane are found by rotating $2\theta = 120°$ from the stress point σ_1, as shown in Figure 1.8. The shear stress on this plane can be found from Figure 1.8 by direct measurement, or by observing that

$$\tau_{60°} = \frac{1}{2}(\sigma_1 - \sigma_2)\sin 120°$$

$$= 0.866 \text{ MPa}$$

1.6 Three-dimensional stress

In the body of a stressed material, the three-dimensional stresses at a point can be represented as if acting on a small cubical element of the material, as shown in Figure 1.9. The nine stresses shown can be set out in an orderly array (a matrix) called the stress tensor:

$$I_\sigma = \begin{bmatrix} \sigma_z & \tau_{zy} & \tau_{zx} \\ \tau_{yz} & \sigma_y & \tau_{yx} \\ \tau_{xz} & \tau_{xy} & \sigma_x \end{bmatrix} \tag{1.24}$$

where the σ terms are the normal stresses and the τ terms are the shear stresses acting on the faces of the element.

It was seen in Section 1.3.2 that $\tau_{zy} = -\tau_{yz}$ and, by the same reasoning, it follows that only six of the terms in equation 1.24 are independent. These are the terms σ_x, σ_y, σ_z, τ_{xy}, τ_{yz}, τ_{zx}, and the matrix is symmetrical about the diagonal containing the normal stresses.

If the stress state in the body remains the same, but the reference axes are rotated to the directions 1, 2, 3 coincident with the principal stress directions, the stress tensor becomes

$$I_\sigma = \begin{bmatrix} \sigma_1 & 0 & 0 \\ 0 & \sigma_2 & 0 \\ 0 & 0 & \sigma_3 \end{bmatrix} \tag{1.25}$$

By adopting an approach identical to that in Section 1.3.2 and applying it to the three 'two-dimensional' directions in Figure 1.9, i.e. $z-y$, $y-x$, $x-z$, it is possible to set up three equations corresponding to equation 1.12 (one of them identical to equation 1.12) and three equations corresponding to equation 1.13 (one of them identical to equation 1.13).

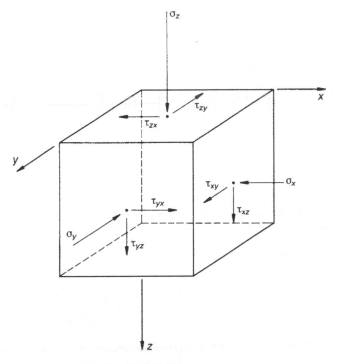

Figure 1.9 Three-dimensional stresses on a cubical element.

No simple method exists for drawing Mohr circles to represent the general case, in which both normal and shear stresses act on all six faces of the cubical element in Figure 1.9. There are two simple cases, however, which can be represented by three Mohr circles:

1. A cubical element which has only normal stresses (i.e. principal stresses) acting on the six faces.
2. A cubical element which has only normal stress (a principal stress) acting on one pair of opposite parallel faces, but has both normal and shear stresses acting on both the remaining pairs of faces.

An example of case 1 is shown in Figure 1.10(a), which depicts a cubical element with compressive principal stresses acting on its six faces. The corresponding three Mohr stress diagrams are shown in Figure 1.10(b). It can be proved that stress conditions on any plane within the element must fall within the shaded area, but it is usually sufficient to be able to determine stresses on planes which are perpendicular to at least one opposite pair of element boundary faces. Stresses on these planes lie on the circles bounding the shaded areas.

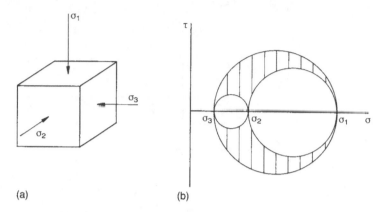

(a)   (b)

Figure 1.10 Cubical element with principal stresses only acting on its faces: (a) stresses; (b) Mohr circles.

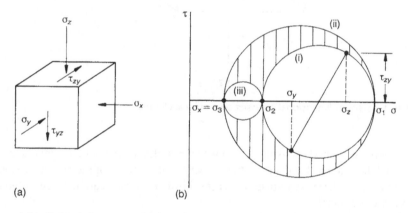

(a)   (b)

Figure 1.11 Cubical element with normal stresses on all faces and shear stresses on two pairs of opposite faces: (a) stresses; (b) Mohr circles.

An example of case 2 is shown in Figure 1.11, which depicts a cubical element with compressive normal stresses acting on all six faces and shear stresses on two pairs of opposite faces. Again, in this case, stresses on all planes within the element lie within the shaded area, with stresses on all planes which are perpendicular to at least one pair of element faces lying on one of the boundary circles. The sequence of drawing these circles consists firstly of locating stress points σ_z, τ_{zy} and σ_y, τ_{yz}, then drawing circle (i) through these with its centre on the $\tau = 0$ axis. This locates the principal stresses σ_1 and σ_2. As the third principal stress is known, circles (ii) and (iii) can now be drawn. In the case shown $\sigma_1 > \sigma_2 > \sigma_3$. In geotechnics it is conventional to adopt this terminology, interrelating the three principal stresses:

1. σ_1 is the **major** principal stress;
2. σ_2 is the **intermediate** principal stress;
3. σ_3 is the **minor** principal stress.

EXAMPLE 1.4 MOHR CIRCLES FOR THREE-DIMENSIONAL STRESSES

A piece of sandstone is cut into the shape of a cube with 100 mm long edges. Forces of 5 kN, 10 kN and 20 kN, respectively, act uniformly on, and normal to, the three pairs of faces of the cube. Evaluate the major, intermediate and minor principal stresses in the rock and draw the Mohr circles of stress. What is the maximum shear stress in the rock, and the orientation of the planes on which it acts?

Solution

As the area of each face of the cube equals 0.01 m², the three principal stresses are:
1. major principal stress,

$$\sigma_1 = \frac{20}{0.01} \times 10^{-3} = 2.0 \text{ MPa}$$

2. intermediate principal stress,

$$\sigma_2 = \frac{10}{0.01} \times 10^{-3} = 1.0 \text{ MPa}$$

3. minor principal stress,

$$\sigma_3 = \frac{5}{0.01} \times 10^{-3} = 0.5 \text{ MPa}$$

The directions of these stresses are shown in Figure 1.12(a) and the resulting Mohr stress circles in Figure 1.12(b).

The maximum shear stress in the rock is equal to the radius of the largest Mohr circle, i.e.

$$\tau_{max} = \frac{1}{2}(\sigma_1 - \sigma_3) = 0.75 \text{ MPa}$$

As the point of maximum shear stress is reached in Figure 1.12(b) by rotating either 90° anticlockwise from σ_1 or 90° clockwise from σ_3, the planes on which

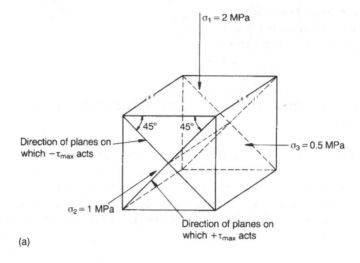

(a)

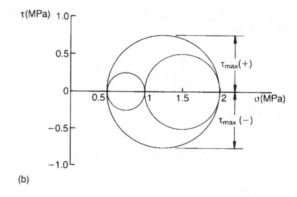

(b)

Figure 1.12 **Example 1.4.**

$+\tau_{max}$ acts are inclined at 45° to the boundary faces on which σ_1 and σ_3 act and are perpendicular to the faces on which σ_2 acts, as shown in Figure 1.12(a). Shear stresses of the same magnitude, but negative according to the sign conventions used in geotechnics, act on planes orthogonal to these as shown in Figure 1.12(b).

EXAMPLE 1.5 PLANE STRAIN – APPLIED PRINCIPAL STRESSES

A test specimen of stiff clay, cut into the shape of a cube, is constrained in one direction by immovable smooth (frictionless) plates acting on opposite vertical faces as shown in Figure 1.13(a). Principal stresses σ_1 and σ_3 act on the other pairs of faces. If the shear strength of the clay is 100 kPa and the specimen is subjected to compression under increasing σ_1, holding σ_3 constant at 80 kPa, find σ_1 at failure. Draw Mohr circles at failure, assuming

$$\sigma_2 = 0.4\left(\sigma_1 + \sigma_3\right)$$

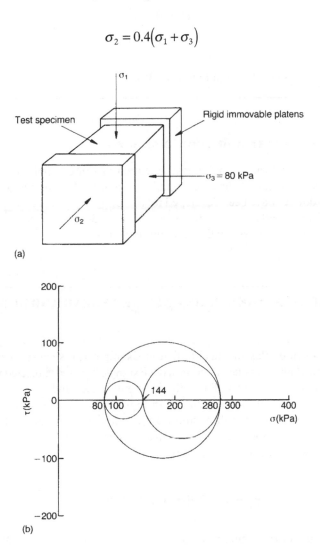

(a)

(b)

Figure 1.13 Example 1.5.

Solution

At failure,

$$\tau_{max} = \tfrac{1}{2}\left(\sigma_1 - \sigma_3\right)$$
$$\therefore \quad 100 = \tfrac{1}{2}\left(\sigma_1 - 80\right)$$
$$\therefore \quad \sigma_1 = 280 \text{ kPa}$$
$$\sigma_2 = 0.4\left(\sigma_1 + \sigma_3\right)$$
$$= 144 \text{ kPa}$$

The Mohr circles are shown in Figure 1.13(b).

1.7 Direct shear and simple shear

It is possible to deform and fail a test specimen of soil by applying shear stress only, or a combination of shear stress and normal stress. A number of laboratory soil testing devices have been developed for this purpose. Before considering these it is useful to consider the specimen in Figure 1.13(a), taken to failure by the application of shear stresses rather than normal stresses.

EXAMPLE 1.6 PLANE STRAIN – APPLIED SHEAR AND NORMAL STRESSES

A test specimen of clay identical to that in Example 1.5 is constrained between smooth parallel plates in one direction, as in Example 1.5, and subjected to constant uniform pressures $\sigma_z = 150$ kPa on the horizontal faces and $\sigma_x = 90$ kPa on the unconstrained vertical faces, as shown in Figure 1.14(a). The specimen is then taken to failure by applying shear stresses τ_{zx}, τ_{xz} to the unconstrained faces. Draw the Mohr circles for the stresses in the specimen at failure, assuming the normal stress on the smooth platens is given by

$$\sigma_y = \sigma_2 = 0.4\left(\sigma_z + \sigma_x\right) = 0.4\left(\sigma_1 + \sigma_3\right)$$

Find the values of σ_1, σ_3 and τ_{zx} at failure, and the inclination to the horizontal of the planes on which the principal stresses act.

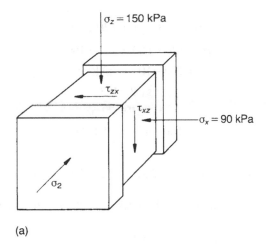

(a)

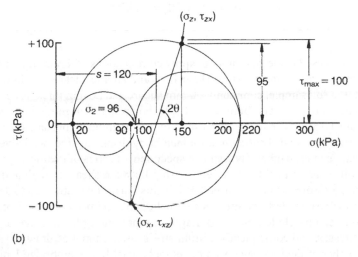

(b)

Figure 1.14 Example 1.6.

Solution

Plot $\sigma_z = 150$ kPa and $\sigma_x = 90$ kPa on the $\tau = 0$ axis, as shown in Figure 1.14(b). The centre of the Mohr stress circles for z–x stresses lies midway between σ_z and σ_x at $s = 120$ kPa, and the circle has a radius of 100 kPa, equal to the shear strength of the clay. The stress points (σ_z, τ_{zx}) and (σ_x, τ_{xz}) can be found by projecting vertically from σ_z and σ_x respectively to intersect the stress circle, noting in Figure 1.14(a) that τ_{zx} is positive and τ_{xz} is negative.

It can be seen from Figure 1.14(b) that

$$s = \tfrac{1}{2}\left(\sigma_z + \sigma_x\right) = 120 \text{ kPa}$$
$$\sigma_1 = 120 + 100 = 220 \text{ kPa}$$
$$\sigma_3 = 120 - 100 = 20 \text{ kPa}$$
$$\cos 2\theta = 0.3$$
$$\therefore \quad 2\theta = 72.5°$$
$$\tau_{zx} = 100 \sin 2\theta = 95 \text{ kPa}$$

Thus σ_1 acts on planes inclined at 36.3° to the horizontal and σ_3 on planes orthogonal to this.

The remaining stress circles can be drawn as shown in Figure 1.14(b), by evaluating σ_2:

$$\sigma_2 = 0.4\left(220 + 20\right) = 96 \text{ kPa}$$

The devices used in the laboratory to apply combinations of direct stress and shear stress are the direct shear test, or shear box, and simple shear tests. In the direct shear test a soil sample, commonly 60 mm square and 20 mm thick, is confined in a stiff box, open top and bottom, and split horizontally at midheight as shown in Figure 1.15(a). The test is performed by maintaining a constant vertical load on the specimen and shearing it at midheight by displacing the two halves of the box relative to each other. By testing a specimen, or series of identical specimens, under different vertical loads, a relationship between shear strength and normal (vertical) pressure can be established. While this is a useful routine test to determine soil strength parameters, the stresses within the soil specimen are not uniform. Only one point can be plotted on a τ–σ diagram, representing the average direct and shear stresses, and consequently a Mohr stress circle cannot be drawn.

A number of devices have been developed, notably at Cambridge University, to apply simple shear to soil specimens, in the manner illustrated in Example 1.6. The Cambridge devices use rectangular test specimens, two opposite vertical faces of which are constrained by immovable rigid platens to give plane strain (i.e. zero strain) conditions in one direction. The other two vertical faces are held against rigid platens, which rotate during shear, as shown in Figure 1.15(b). Normal stresses and shear stresses are applied across the horizontal faces. Shear stress across the specimen is not uniform, tending towards a maximum value in the middle and low values at the ends, as shown in Figure 1.15(c), because of the difficulty of satisfying the equality of conjugate horizontal and vertical stresses on the boundaries of the test specimen. Load cells built into these devices, to measure shear and normal forces, have indicated that the middle third of the specimen does deform reasonably well in pure shear (Roscoe, 1970). These measurements may

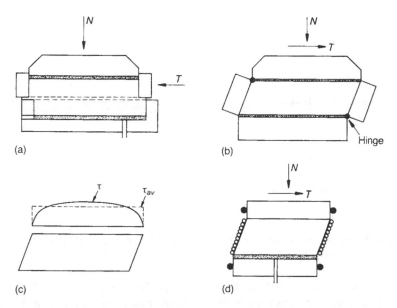

Figure 1.15 Laboratory shear tests: (a) direct shear; (b) Cambridge simple shear device; (c) distribution of horizontal shear stress in simple shear device; (d) NGI simple shear device.

also allow Mohr stress circles to be drawn. While some useful research data have been obtained from these simple shear devices (e.g. Stroud, 1971), they are not suitable for routine laboratory testing.

A device developed at the Norwegian Geotechnical Institute uses a circular test specimen confined in a rubber sleeve reinforced by spiral wire, as shown in Figure 1.15(d). Normal force N and shear force T are applied to the top surface of the specimen as shown. As shear stresses on vertical faces are negligible, the conjugate shear stresses on the horizontal faces near the edges must also be negligible. One stress point on a τ–σ diagram can be plotted, assuming average σ_z and τ_{zx} values, but the stress circle cannot be completed as no other stresses are known. Nevertheless the test is useful from a practical point of view as it does give a measure of shear strength, and its simplicity allows it to be used as a routine laboratory test.

1.8 Triaxial stress

The triaxial test is the most commonly used laboratory test to determine the strength parameters for soils and rocks. As this test, and its interpretation, are dealt with at some length in Chapter 2, it is appropriate here only to examine briefly the basic stress conditions.

It has the advantage that an easily trimmed cylindrical test specimen is used, usually with a length to diameter ratio of 2.0. Diameters of test specimens are

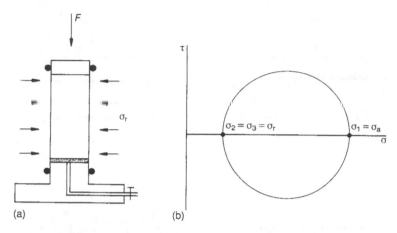

Figure 1.16 Triaxial test: (a) radial stress σ_r and applied axial force F; (b) Mohr stress circle ($\sigma_a = F/A$, where A is the area of the specimen).

commonly in the range of 38 mm to 100 mm. As illustrated in Figure 1.16(a), the test is usually performed by maintaining a constant radial stress σ_r, while increasing the axial force F to fail the specimen in compression. Thus, the axial stress σ_a is the major principal stress σ_1, and the intermediate and minor principal stresses σ_2 and σ_3 are both equal to the radial stress σ_r. All stresses in the test specimen are known and, as $\sigma_2 = \sigma_3$, they are fully represented by a single stress circle as shown in Figure 1.16(b).

EXAMPLE 1.7 TRIAXIAL STRESSES

Draw the Mohr stress circle at failure for a triaxial compression test on a specimen of stiff clay with a shear strength of 100 kPa, if the radial stress is maintained constant at 80 kPa. Find the inclination θ to the radial direction of the planes on which the shear stress is one-half the maximum shear stress, and determine the normal stresses acting on these planes.

Solution

$$\sigma_2 = \sigma_3 = 80 \text{ kPa}$$

As $\tau_{max} = 100$ kPa, the diameter of the stress circle at failure, as shown in Figure 1.17, is 200 kPa. Thus

$$\sigma_1 = \sigma_3 + 200 = 280 \text{ kPa}$$

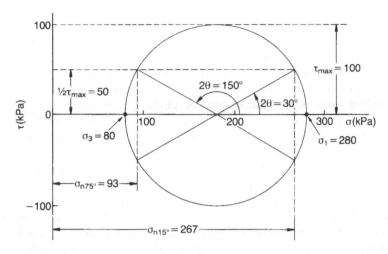

Figure 1.17 Example 1.7.

The angle 2θ is given by

$$\sin 2\theta = \frac{\tau_{max}}{2\tau_{max}} = 0.5$$

$$\therefore \quad \theta = 15° \text{ or } 75°$$

The normal stresses $\sigma_{n15°}$ and $\sigma_{n75°}$ acting on these planes can be found readily from Figure 1.17.

$$\sigma_{n15°} = 180 + 100\cos 2\theta$$
$$= 267 \text{ kPa}$$
$$\sigma_{n75°} = 180 - 100\cos 2\theta$$
$$= 93 \text{ kPa}$$

Note. In considering relative stress orientations it is possible to work with the stress directions themselves or the direction of the planes on which the stresses act. In Figure 1.17 the angle between the radial (horizontal) planes on which σ_1 acts and the planes on which $\sigma_{n15°}$ acts is 15°, which is also the angle between directions of σ_1 and $\sigma_{n15°}$. Similarly $\sigma_{n75°}$ acts on planes inclined at 75° to the radial planes, which is also the angle between the directions of σ_1 and $\sigma_{n75°}$.

1.9 Pole points

As shown in Sections 1.3 and 1.4, if the stresses (σ_z, τ_{zx}), (σ_x, τ_{xz}) on orthogonal planes are known, the stresses $(\sigma_{n\theta}, \tau_\theta)$ on any other plane at an angle θ can be found by using equations 1.12 and 1.13 or, graphically, by rotating the stress point on the Mohr circle by 2θ. A simple alternative method can also be used, by establishing a pole point on the Mohr stress circle.

Two pole points can be established, one relating to the directions of action of the stresses and the other relating to the directions of the planes on which the stresses are acting.

Referring to Figure 1.18, the pole point P_s for stresses is found either by projecting a line from the stress point (σ_z, τ_{zx}) in the direction of action of σ_z, i.e. vertically, until it intersects the stress circle; or by projecting a line from the stress point (σ_z, τ_{xz}) in the direction of action of σ_x, i.e. horizontally, until it intersects the stress circle. Either projection will give the unique pole point P_s. The pole point P_p for planes is found either by projecting a line from the stress point (σ_z, τ_{zx}) in the direction of the plane on which these stresses are acting, i.e. horizontally; or by projecting vertically from the (σ_x, τ_{xz}) stress point.

Although either pole point can be used with equal facility it is usual to work with the pole point for planes P_p. It is this pole point which is used throughout this text. The use of the pole point P_p to locate stresses (σ_c, τ_{ca}) at an angle θ to the reference stress (σ_z, τ_{zx}) is shown in Figure 1.19. The stress point on the Mohr circle is found by simply projecting a line from P_p parallel to the plane on which (σ_c, τ_{ca}) acts until it intersects the circle at point D.

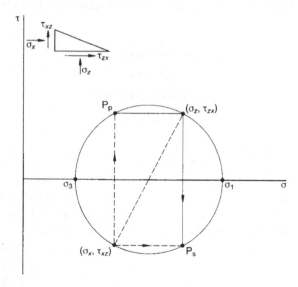

Figure 1.18 Pole points P_s for stress directions and P_p for directions of planes on which stresses act.

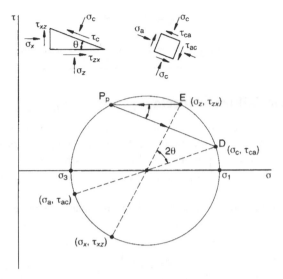

Figure 1.19 Use of pole point P to locate stresses (σ_c, τ_{ca}) at angle θ to the reference stress direction.

EXAMPLE 1.8 USE OF POLE POINT

Find the angle θ in Example 1.7 between the radial planes and planes on which $\tau = \frac{1}{2}\tau_{max}$, using the pole point method.

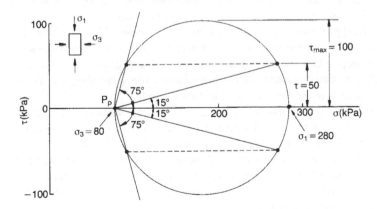

Figure 1.20 Example 1.8.

Solution

The solution is shown in Figure 1.20, from which it can be seen that

$$\theta = 15° \text{ or } 75°$$

The values of $\sigma_{n15°}$ and $\sigma_{n75°}$ can easily be evaluated using this diagram.

1.10 Basic failure criteria

In 1773 Coulomb read his paper to the French Academy of Sciences, which dealt with a variety of matters ranging from the strength of beams and the stability of arches, to earth pressures and the shear strength of masonry and soils. After being refereed by his peers the paper was published by the Academy in 1776.

Coulomb proposed for masonry and soil a shear resistance expression of the form

$$S = ca + \frac{1}{n} N \qquad (1.26)$$

where c is the (non-directional) cohesion per unit area; a is the area of the shear plane; N is the normal force on the shear plane; $1/n$ is the coefficient of internal friction.

In modern terms $n = \cot \phi$ and equation 1.26 is usually written in the form

$$\tau_f = c + \sigma_n \tan \phi \qquad (1.27)$$

where τ_f is the shear strength per unit area; c is the unit cohesion; σ_n is the normal stress on the shear plane; ϕ is the angle of shearing resistance.

For all practical purposes at least, the validity of equation 1.27 for soils is now universally accepted, but parameters c and ϕ may take many different values for the same soil, depending on stress path, stress level and drainage conditions. Equation 1.27 can also be applied in rock mechanics for shear along joints and discontinuities and in some cases to the intact rock itself.

At the time Mohr was working on the graphical representation of stress at a point, most engineers concerned with stress analysis favoured the maximum strain theory of Saint-Venant as their failure criterion. Aware of the fact that this criterion did not give good agreement with experiments on steel specimens, Mohr promoted the use of a failure criterion based on limiting shear resistance, and furthermore proposed that stress circles should be drawn to give a full understanding of stress conditions at failure.

As an illustration of the shear stress criterion, Mohr used the example of cast iron tested to failure in compression (failure stress σ_c), in tension (failure stress σ_t) and in pure shear (failure stress τ_{ult}). He drew the failure envelopes as shown in Figure 1.21, just touching the compression and tension circles, and showed that the failure stress in shear could then be deduced by drawing a circle with its

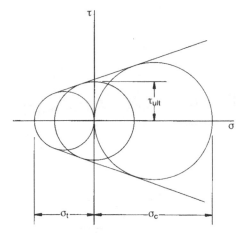

Figure 1.21 Mohr's stress circles and strength envelopes for cast iron.

centre at the origin of stresses and with radius τ_{ult} such that the circle just touched the two envelopes. From this construction:

$$\tau_{ult} = \frac{\sigma_c \sigma_t}{\sigma_c + \sigma_t} \qquad (1.28)$$

which agreed satisfactorily with experiments.

Although the motivations of Coulomb and Mohr towards developing a failure criterion were very different, and on different materials, the end point was much the same: a stress-dependent criterion based on shear resistance, familiar to geotechnical engineers as the 'Mohr–Coulomb criterion'.

The combination of the Mohr stress circle with the Mohr–Coulomb failure criterion not only gives a valuable understanding of stress conditions at failure, but also provides a very powerful tool in geotechnical analyses.

Referring to Figure 1.22 which shows a Mohr stress circle for the typical case of a cylindrical triaxial specimen tested to failure, in axial compression, exhibiting strength characteristics c, ϕ, the following useful expressions can be deduced:

$$\frac{\sigma_1}{\sigma_3} = \frac{2c\cos\phi}{\sigma_3(1-\sin\phi)} + \frac{1+\sin\phi}{1-\sin\phi} \qquad (1.29)$$

In the special case where $c = 0$:

$$\frac{\sigma_1}{\sigma_3} = \frac{1+\sin\phi}{1-\sin\phi} \qquad (1.30a)$$

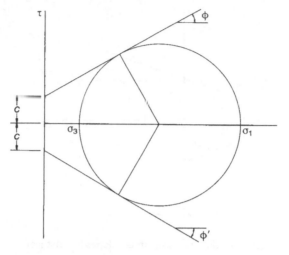

Figure 1.22 Shear strength parameters c and ϕ.

or

$$\frac{\sigma_1}{\sigma_3} = \tan^2\left(\frac{\pi}{4} + \frac{\phi}{2}\right) \qquad (1.30b)$$

In the special case where $\phi = 0$:

$$\left(\sigma_1 - \sigma_3\right) = 2c \qquad (1.31)$$

In plotting the results from conventional laboratory triaxial tests, in which the test specimens are compressed axially, it is usual to plot only the semicircle and envelope above the $\tau = 0$ axis in Figure 1.22. A method of deducing the best-fit envelope from experimental data using the least squares method has been described by Balmer (1952).

1.11 Effective stress and stress history

It was the establishment of the principle of effective stress by Terzaghi (1936) which placed soil mechanics on a firm scientific basis. The strength and defor-mation characteristics of a soil are determined primarily by the magnitude of effective stress and the stress history which the soil has experienced. The effective stresses in a soil are those stresses carried by the soil skeleton, through interparticle contact forces. Referring to Figure 1.23, if the soil voids are filled with water, the total normal stress σ across any interface within the soil mass is the sum of the

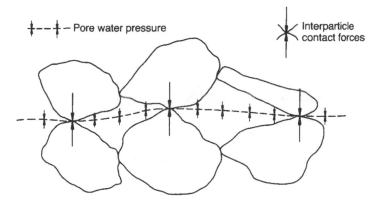

+--+-- Pore water pressure

Interparticle contact forces

Figure 1.23 Interparticle contact forces and pore water pressure within a soil mass.

pressure u in the pore water and the effective stress σ' carried by the particle contacts, i.e.

$$\sigma = \sigma' + u \qquad (1.32)$$

This expression embodies the assumption of point contacts between the particles (and hence 'zero' area of contact), so that the pore water pressure acts over the whole area of any interface passing through particle contact points, as shown in Figure 1.23.

During deposition of a soil under water in the field, the vertical effective stress at any point in the soil mass increases as the depth of overlying soil increases. This causes the soil to consolidate into a closer packing, following a curve of the voids ratio e against consolidation pressure σ' such as AB in Figure 1.24. The voids ratio is the ratio of volume of soil voids to volume of soil particles. It is common practice to plot e against log σ', in which case AB often approximates to a straight line.

If after reaching point B in Figure 1.24 some of the overburden is removed, for example by erosion, the soil follows a swelling curve such as BC in Figure 1.24. At point B the soil is normally consolidated and at point C it is over-consolidated. The overconsolidation ratio (OCR) is given by the expression

$$OCR = \frac{\sigma'_B}{\sigma'_C} \qquad (1.33)$$

That is, OCR is the ratio of maximum vertical effective stress which the soil has experienced in its past history σ'_{vmax} to the existing vertical effective stress σ'_{v0} and can thus be expressed as

$$OCR = \frac{\sigma'_{vmax}}{\sigma'_{v0}}$$

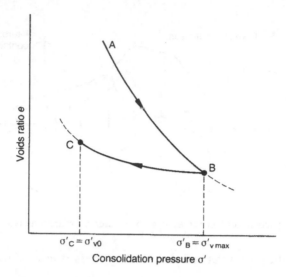

Figure 1.24 Normal or virgin consolidation (AB) and swelling (BC).

Soils are often referred to as 'lightly overconsolidated' or 'heavily overconsolidated'. The approximate values of OCR corresponding to such qualitative descriptions are shown in the following table.

Description	OCR
Normally consolidated	1
Lightly overconsolidated	1–3
Moderately overconsolidated	3–9
Heavily overconsolidated	>9

For research purposes a predetermined OCR is often achieved in the laboratory by reconstituting soil at a high water content and then submitting it to an appropriate consolidation–swelling cycle.

1.12 Mohr strain circle

As with stresses, consideration here will be confined to two-dimensional strain states in the z, x plane.

The applied stresses in Figure 1.25(a) cause the distortions of the element OABC shown in Figure 1.25(b), where:

1. $\varepsilon_z, \varepsilon_x$ are direct strains;
2. $\varepsilon_{zx}, \varepsilon_{xz}$ are pure shear strains.

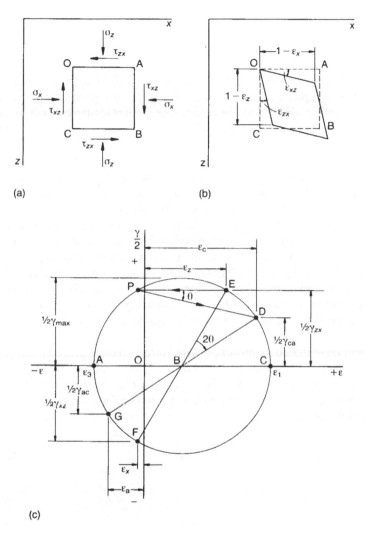

Figure 1.25 Two-dimensional strains: (a) stresses on element OABC; (b) resulting strains; (c) Mohr strain circle.

It is important to note that engineers usually work in terms of the engineers' shear strain γ, which is the total shear distortion of the element. Thus

$$\gamma = \varepsilon_{zx} + \varepsilon_{xz} \tag{1.34}$$

in magnitude, but

$$\varepsilon_{zx} = \varepsilon_{xz} \tag{1.35}$$

Thus

$$\gamma = 2\varepsilon_{zx} \qquad (1.36)$$

If the strains in element OABC in Figure 1.25(b) caused by applied stresses (σ_z, σ_y, τ_{zx}) shown in Figure 1.25(a) are known, these can be represented graphically on a Mohr circle of strain. The strains in any other element rotated through an angle θ can then be found in a manner identical to that for finding stresses using a stress circle.

Figure 1.25(c) shows the Mohr circle of strain plotted for the known strains in element OABC. It will be noted that pure shear strain $\gamma/2$ is plotted against principal strains in this diagram. The strains shown in Figure 1.25(b) plot as points E and F in Figure 1.25(c). Strains in an element with orthogonal directions a, c at an angle θ to element OABC can be found either by rotating the strain point on the circle from E to D by 2θ or by establishing the pole point P, then projecting the line PE at an angle θ from P which again gives the required strain point.

In Figure 1.25(c) the pole point P has been established by projecting a line from E in a direction normal to the direction of strain (giving a pole point corresponding to the pole point for planes in the Mohr stress circle). As with the stress circle, a pole point can also be established by projecting from D in the direction of the strain ε_z.

The magnitudes and directions of principal direct strains ε_1, ε_3 and maximum shear strain $(\gamma/2)_{max}$ can readily be found from Figure 1.25(c). It is also possible to find the volumetric strain ε_V from Figure 1.25(c), if plane strain conditions are assumed, that is the direct strain in a direction perpendicular to the xz plane is zero. Then for small strains,

$$\varepsilon_V = \left(\varepsilon_1 + \varepsilon_3\right) \qquad (1.37)$$

but

$$BC = \tfrac{1}{2}\left(\varepsilon_1 - \varepsilon_3\right)$$
$$OB = \left(\varepsilon_1 - BC\right)$$
$$\therefore OB = \tfrac{1}{2}\left(\varepsilon_1 + \varepsilon_3\right)$$

thus

$$OB = \frac{\varepsilon_V}{2} \qquad (1.38)$$

1.13 Angle of dilatancy

Volume changes which occur during shearing of soils can be conveniently characterized by the dilatancy angle ψ, given by

$$\sin \psi = -\frac{\delta \varepsilon_{\mathrm{v}}}{\delta \gamma_{\mathrm{max}}} \tag{1.39}$$

where $\delta \varepsilon_{\mathrm{v}}$ is a small increment of volumetric strain and $\delta \gamma_{\mathrm{max}}$ is the corresponding small increment of maximum shear strain. The negative sign is needed to reconcile the anomalous convention that compressive strains are positive, whereas ψ is positive for volumetric increases. Thus

$$\sin \psi = -\frac{\delta \varepsilon_1 + \delta \varepsilon_3}{\delta \varepsilon_1 - \delta \varepsilon_3} \tag{1.40a}$$

or

$$\sin \psi = -\frac{\left(\delta \varepsilon_1 / \delta \varepsilon_3\right) + 1}{\left(\delta \varepsilon_1 / \delta \varepsilon_3\right) - 1} \tag{1.40b}$$

The dilatancy angle ψ for plane strain deformation is shown on a Mohr strain diagram in Figure 1.26. As the negative principal strain increment $\delta \varepsilon_3$ in this example exceeds in magnitude the positive principal strain increment $\delta \varepsilon_1$, the volume of the soil is increasing and ψ is positive.

Points A and B on the strain circle in Figure 1.26 represent conditions of zero direct strain increment, i.e. $\delta \varepsilon_{\mathrm{n}} = 0$. The directions in which $\delta \varepsilon_{\mathrm{n}} = 0$ are known as zero extension lines, and their orientations may be found as follows, referring to Figure 1.26.

1. Assuming the principal strain increment $\delta \varepsilon_1$ to be acting vertically, draw a horizontal line from $\delta \varepsilon_1$ to locate the pole point P on the strain circle.
2. Lines PA and PB represent the directions of planes perpendicular to the directions of zero strains $\delta \varepsilon_{\mathrm{n}} = 0$.
3. Points C and D are located by drawing horizontal lines from A and B respectively to meet the strain circle; lines PC and PD are then the directions of the lines of zero extension.

Note that a simpler diagram to determine the directions of zero extension lines will result if the pole point for strain directions (rather than normal to strain directions) is used. In this case point E is the pole point and EA, EB are the directions of zero extension lines (see for example Figure 8.19).

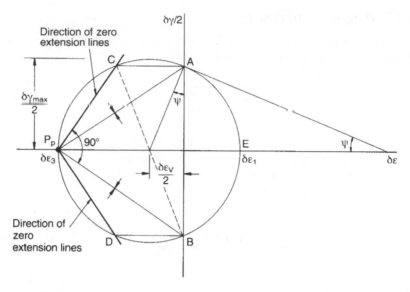

Figure 1.26 Mohr strain circle showing dilatancy angle ψ and volumetric strain increment $\delta\varepsilon_v$.

EXAMPLE 1.9 DETERMINATION OF STRAINS AND ZERO EXTENSION LINES

Figure 1.27(a) shows an element of soil behind a vertical retaining wall, which is supported by horizontal props. If a slight yielding of the props allows the element to strain horizontally by an amount $\delta\varepsilon_h = 0.16\%$, find the vertical strain $\delta\varepsilon_v$ if the angle of dilation for the soil is +15°. Assume $\delta\varepsilon_h$ and $\delta\varepsilon_v$ are principal strains. Determine the volumetric strain $\delta\varepsilon_v$ of the element and the shear strains γ_t across zero extension lines. Find the orientation of the zero extension lines. Assume plane strain conditions in the longitudinal direction of the retaining wall.

Solution

Referring to Figure 1.27(b), the radius r of the strain circle is given by:

$$r + r\sin 15° = 0.16\%$$
$$\therefore \quad r = 0.13\%$$
$$\delta\varepsilon_v = \delta\varepsilon_1 = r - r\sin 15°$$
$$= 0.09\%$$

Volumetric strain $\delta\varepsilon_V = -2r\sin 15°$
$$= -0.066\%$$

Shear strain across zero extension lines $= 2r\cos 15°$
$$= -0.25\%$$

Orientation of zero extension lines $\tan\alpha = \tan\beta = \dfrac{r\cos 15°}{r(1-\sin 15°)}$
$$= 52.5°$$

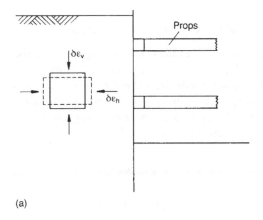

(a)

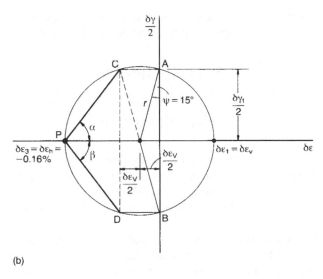

(b)

Figure 1.27 Example 1.9.

Chapter 2

Failure states in soil

2.1 Total and effective stress circles

As stated in Section 1.11, and illustrated in Figure 1.23, the total stress across any interface through particle contact points in a soil mass is made up of the pore water pressure in the soil voids and the effective stress, equal to the summation of interparticle forces over a unit area. This leads to the effective stress equation given by equation 1.32. As the pore water pressure acts with equal intensity in all directions, it follows from equation 1.32 that

$$\sigma_1 = \sigma_1' + u \tag{2.1a}$$

$$\sigma_3 = \sigma_3' + u \tag{2.1b}$$

where σ_1, σ_3 are the major and minor total principal stresses; σ_1', σ_3' are the major and minor effective principal stresses. From equations 2.1a and 2.1b,

$$\tfrac{1}{2}\left(\sigma_1' + \sigma_3'\right) = \tfrac{1}{2}\left(\sigma_1 + \sigma_3\right) - u \tag{2.2a}$$

$$\tfrac{1}{2}\left(\sigma_1' - \sigma_3'\right) = \tfrac{1}{2}\left(\sigma_1 - \sigma_3\right) \tag{2.2b}$$

It follows from equations 2.2a and 2.2b that the total stress and effective stress Mohr circles have the same radius, but are separated along the σ axis by an amount equal to the pore pressure. This is shown for positive (compression) pore pressure $+u$ in Figure 2.1(a) and for negative (tension) pore pressure $-u$ in Figure 2.1(b).

It can also be seen from equation 2.2b, and Figures 2.1(a) and 2.1(b), that shear stresses are not affected by pore pressure; that is, the values are identical whether expressed in terms of total stress or effective stress. The physical explanation for this is the inability of the pore water to resist shear stress, so that shear stresses are resisted entirely by contact forces between soil grains.

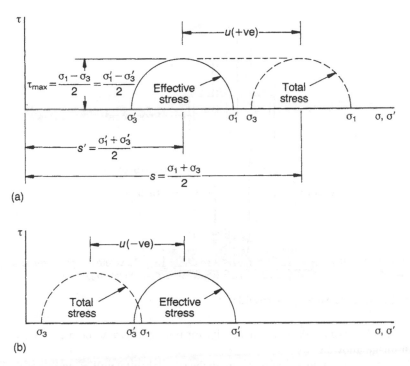

(a)

(b)

Figure 2.1 Relationship between total and effective stress circles if: (a) pore pressure is positive; (b) pore pressure is negative.

2.2 The triaxial test

The Mohr–Coulomb criterion usually fits experimentally determined failure states for soils very well. However, the parameters c and ϕ in equation 1.27 are greatly influenced by test and loading conditions. It is convenient to consider the influence of some of these factors on failure states as measured in the triaxial test in the laboratory. The advantage of this test, for this purpose, lies in the fact that all stresses in the specimen are known and the intermediate principal stress σ_2 must be equal to either the minor principal stress σ_3 or the major principal stress σ_1; and thus stress conditions in the test specimen can be represented fully by the two-dimensional Mohr stress circle. It has the further advantage that it is the most common soil mechanics strength test performed in the laboratory. For a full account of the triaxial test the reader is referred to Bishop and Henkel (1962).

The basic elements of the triaxial test are shown in Figure 2.2. The cylindrical test specimen is placed on a pedestal inside a perspex cell which is filled with water. A latex sheath surrounds the test specimen and is sealed against the base pedestal and the top cap, through which the sample is loaded axially, by means of a ram passing through a bushing in the top of the cell. A duct passing through the base pedestal allows water to drain from the voids of the sample, the volume of

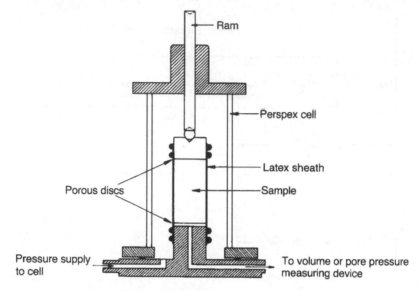

Figure 2.2 Basic elements of the triaxial test.

which can be measured; or, alternatively, for pore pressure to be measured with no drainage allowed.

When the test specimen is first set up in the triaxial cell it will be under zero cell pressure, but in the case of fine-grained saturated soils the water in the soil voids will be under negative pressure, which will be holding the specimen intact and giving it some strength. As the negative pore pressure depends on the development of surface tensions at the air–soil boundary before it is placed in the cell, coarse-grained soils, which cannot sustain these negative pore pressures, have to be set up in a supporting mould until a negative pressure can be applied through the drainage duct in the base.

Initially, then:

$$u = u_e(\text{negative}) \tag{2.3a}$$

$$\sigma_a = \sigma_r = 0 \tag{2.3b}$$

$$\sigma'_a = \sigma'_r = -u_e \tag{2.3c}$$

where σ_a is the axial stress; σ_r is the radial stress (Figure 2.3).

If the drainage valve is kept closed and the cell pressure increased to σ_{cp}, this applied isotropic pressure is taken entirely by the pore water, which is relatively incompressible compared with the soil fabric. Thus, the pore pressure increases from u_e to u_i, but there is no change in effective stress. Thus

$$u_i = \sigma_{cp} + u_e \qquad (2.4a)$$

$$\sigma_a = \sigma_r = \sigma_{cp} \qquad (2.4b)$$

$$\sigma'_a = \sigma'_r = -u_e \qquad (2.4c)$$

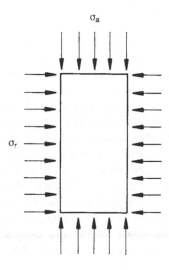

Figure 2.3 Triaxial test stresses: axial stress σ_a, radial stress σ_r.

Equation 2.4a applies for any change σ_{cp} in cell pressure. Thus

$$u - u_e = \Delta\sigma_{cp}$$
$$\text{i.e. } \Delta u = \Delta\sigma_{cp} \qquad (2.5)$$

The triaxial test gives great flexibility with respect to possible stress changes, and pore water drainage conditions, in taking the test specimen to failure. With respect to drainage conditions, one of the following three procedures is usually adopted.

1. **Unconsolidated undrained test (UU):** the specimen is taken to failure with no drainage permitted.
2. **Consolidated undrained test (CU):** the drainage valve is initially opened to allow the pore pressure u_i to dissipate to zero, and then closed so that the specimen is taken to failure without permitting any further drainage. It is common to apply a 'back pressure', that is a positive pore pressure, to the specimen initially, balanced by an equal increment in cell pressure to avoid

changing the effective stress. This is to ensure that any air in the soil voids or in the ducts connecting to the pore pressure measuring device is driven into solution in the water. It also decreases the possibility of cavitation, that is water vapour forming, or air coming out of solution in the water, if large negative changes in pore pressure take place during a test.

3. **Drained test (CD):** the drainage valve is initially opened to allow the pore pressure u_i to dissipate to zero, and is kept open while the specimen is taken to failure at a sufficiently slow rate to allow excess pore pressures to dissipate.

It is possible to take the specimen in the triaxial cell to failure either in axial **compression** or axial **extension**. The specimen can be compressed axially either by increasing the axial compressive stress or decreasing the radial stress (i.e. the cell pressure), or a combination of both. Axial extension can be achieved either by decreasing the axial stress or increasing the radial stress, or a combination of both. It should be noted that all stresses normally remain compressive during extension tests, even when failure is induced by a reduction of axial stress. It is possible for a soil specimen to sustain a tensile total stress, which can be carried by a corresponding negative pore pressure, but it is difficult to effect a seating between the testing device and the soil, which will allow a tensile stress to be applied. Alternatively, a specially shaped specimen can be used (Bishop and Garga, 1969). It is usually assumed that soils are unable to sustain any negative effective stress, although in fact some intact natural soils can sustain very small values which, again, are very difficult to measure (Parry and Nadarajah, 1974b, Bishop and Garga, 1969).

Figure 2.4 shows Mohr circles at failure for drained compression and extension tests on identical specimens under the same initial isotropic stress and taken to failure by increasing and decreasing the axial stress respectively, keeping the radial stress σ_r constant. This diagram assumes $c' = 0$ and ϕ' to be the same for the two different tests, which in many soils will not be the case.

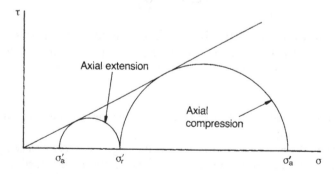

Figure 2.4 Stress circles at failure for drained triaxial compression and extension tests.

EXAMPLE 2.1 PORE PRESSURE IN TRIAXIAL TEST UNDER ISOTROPIC STRESS

An undisturbed cylindrical sample of saturated clay, 38 mm diameter by 76 mm high, is placed in a triaxial cell and put under an all-round cell pressure of 80 kPa, with the drainage valve closed.

1. If the pore pressure measured under this pressure is 50 kPa, what was the value of the pore pressure before applying the cell pressure?
2. If drainage of pore water from the sample is now permitted until the pore pressure drops to zero, how much water will be expelled if the sample changes volume by 0.01% for every increase of 1 kPa in effective stress?
3. If the drainage valve is now closed and the cell pressure decreased to zero, what will be the final pore pressure in the sample?

Solution

1. $\sigma_{cp} = 80$ kPa $u_i = 50$ kPa

$$\text{Equation 2.4a: } 50 = 80 + u_e$$
$$\therefore u_e = -30 \text{ kPa}$$

2. Volume of sample $= \pi \times 19^2 \times 76$ mm^3
$$= 86\ 193 \text{ mm}^3$$

Change in pore pressure $\Delta u = -50$ kPa

Change in effective stress $= -\Delta u = 50$ kPa

$$\text{Volume of water expelled} = 86\ 193 \times 50 \times \frac{0.01}{100}$$
$$= 431 \text{ mm}^3$$

3. $\Delta\sigma_{cp} = -80$ kPa

Equation 2.5: $\Delta u = -80$ kPa

As the initial pore pressure is zero, the pore pressure after removing the cell pressure is therefore $u = -80$ kPa.

2.3 Triaxial compression tests

Unless otherwise stated, reference to the triaxial test in geotechnical literature invariably means the triaxial compression test – performed by holding the cell pressure constant and increasing the axial stress. Full specification of the test requires a statement of the initial applied stress conditions and whether drained or undrained during shear. In some practical applications, such as investigating the stability of cut slopes, it would be more logical to take the specimen to failure in compression by reducing the radial stress; but this high degree of versatility offered by the triaxial test is rarely exploited in practice.

Axial loading in the triaxial test is usually applied under constant rate of strain, but occasionally load control methods are employed. In triaxial compression:

$$\sigma_1 = \sigma_a \tag{2.6a}$$

$$\sigma_2 = \sigma_3 = \sigma_r \tag{2.6b}$$

In Figure 2.5 typical results which might be expected from UU, CU and CD compression tests on a low to moderately sensitive, lightly overconsolidated, natural clay are plotted as deviator stress $(\sigma_a - \sigma_r)$ and pore pressure u, or volumetric strain ε_V, against axial strain ε_a. Depending upon a number of factors such as soil fabric, stress history, applied stress path and drainage conditions, the plotted curve may show a pronounced peak at failure, with a marked drop in strength under continued straining, or a fairly flat peak (p in Figures 2.5(a), (b)), or simply reach a plateau with no reduction in strength with continued straining. A specimen showing a definite peak under undrained conditions might well exhibit a continuing increase in pore pressure, which may be wholly or partly responsible for the post-peak reduction in strength. Eventually a steady-state shearing condition should be reached with constant shear strength and pore pressure under continued straining. Under ideal circumstances, notably uniform straining throughout the test specimen, this ultimate state is the critical state (Schofield and Wroth, 1968).

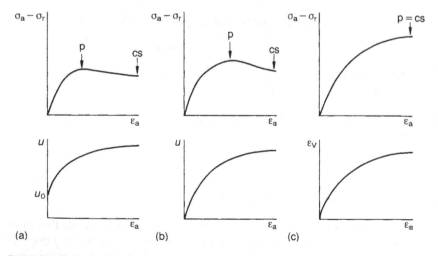

Figure 2.5 Typical stress–strain plots for (a) UU, (b) CU and (c) CD triaxial tests on a lightly overconsolidated natural clay (p, peak; cs, critical state).

In the drained test on lightly overconsolidated clay a plateau rather than a peak failure condition may be reached, because the progressive decrease in volume through the test gives a progressive increase in strength. Ideally, the plateau corresponds to the critical state.

In a heavily overconsolidated state the same soil may exhibit a pronounced peak strength in a drained test, as the specimen may be dilating, and thus weakening as the test progresses. In addition, a pronounced failure plane is likely to form, with the specimen eventually approaching a residual condition (Skempton, 1964) rather than critical state. The limited displacements which can be achieved in the triaxial test do not allow a full residual state to be reached. In the undrained tests a progressive decrease in pore pressure is likely, with progressive increase in strength, and thus a plateau failure might be expected; but in natural clays defects or fissures are often present which cause heavily overconsolidated clays to exhibit a distinct peak in UU and CU tests.

Considering again the typical results for a lightly overconsolidated clay represented in Figure 2.5, the three peak effective stress circles could be reproduced on a single diagram, but an envelope tangential to these three circles would be meaningless, as the peak strength is a transient condition and there is no common factor linking the three circles other than the soil.

Ignoring certain test influences, such as strain rate, an envelope of fundamental importance can be obtained by a tangent to the three critical state circles drawn on one diagram, as in Figure 2.6. It is usually found, and invariably assumed, that the critical state cohesion intercept c'_{cs} is zero. The slope of the envelope is denoted ϕ'_{cs}.

The basis of the critical state concept is that under sustained uniform shearing, specimens of a specific soil achieve unique e vs p'_{cs} and q_{cs} vs p'_{cs} relationships, regardless of the test type and initial soil conditions. This is shown in Figure 2.7. In Figure 2.7(a), A is the initial condition for a lightly overconsolidated soil with characteristics as depicted in Figure 2.5. According to test type, the critical state line is reached at points B (UU test), D (CU test – AC is consolidation phase) and

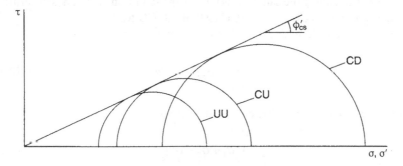

Figure 2.6 Typical stress circles and strength envelope at critical state for UU, CU and CD tests on identical test specimens of a lightly overconsolidated clay.

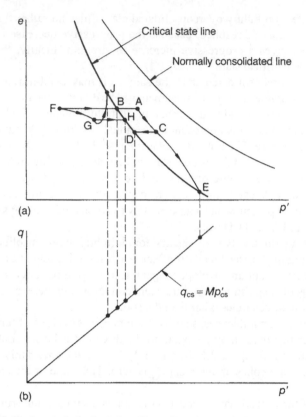

Figure 2.7 Critical state behaviour in soils.

E (CD test). At all points on the critical state line, such as B, D and E, the shear resistance, pore pressure and voids ratio remain constant under continued shearing. A heavily overconsolidated sample with initial conditions represented by point F in Figure 2.7(a), having the same voids ratio as the sample at A, reaches the critical state line at points B (UU test), H (CU test) and J (CD test).

The critical state strength expression is usually written

$$q_{cs} = \left(\sigma_1' - \sigma_3'\right)_{cs} = Mp_{cs}' \tag{2.7}$$

where M is a constant for a particular soil. Putting

$$p_{cs}' = \tfrac{1}{3}\left(\sigma_1' + \sigma_2' + \sigma_3'\right)_{cs} \tag{2.8}$$

equation 2.7 becomes

$$\left(\sigma_1' - \sigma_3'\right)_{cs} = \frac{M}{3}\left(\sigma_1' + \sigma_2' + \sigma_3'\right)_{cs} \tag{2.9}$$

but, for trixial compression, $\sigma_2' = \sigma_3'$ and equation 2.9 reduces to

$$\left(\sigma_1' - \sigma_3'\right)_{cs}' = \frac{M}{3}\left(\sigma_1' + 2\sigma_3'\right)_{cs} \tag{2.10}$$

From equation 1.30a:

$$\left(\frac{\sigma_1'}{\sigma_3'}\right)_{cs} = \frac{1 + \sin\phi_{cs}'}{1 - \sin\phi_{cs}'} \tag{2.11}$$

Combining equations 2.10 and 2.11 gives

$$M = \frac{6\sin\phi_{cs}'}{3 - \sin\phi_{cs}'} \tag{2.12}$$

Two other strength criteria of importance in soils are the Hvorslev (1937) concept and residual strength (Skempton, 1964). The Hvorslev concept is an envelope touching effective stress peak strength circles for specimens of the same soil tested under drained conditions, having the same voids ratio e at failure but different stress conditions. A series of tests with different voids ratios at failure results in a family of parallel envelopes as shown in Figure 2.8 with constant ϕ_e', and c_e' intercepts decreasing with increasing e. This concept has little practical usefulness, partly because of the difficulty of conducting laboratory tests to achieve specific voids ratios at failure and partly because, in the field, various loading possibilities produce different voids ratios.

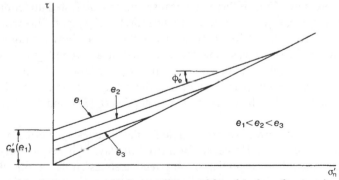

The residual shear strength is usually considered to be of most significance in
Figure 2.8 Hvorslev envelopes.

heavily overconsolidated clays, where a distinct failure plane or discontinuity is usually observed. Under large displacements, strong alignment of clay particles occurs along the failure plane or within a thin failure zone, leading to low values of c'_r, ϕ'_r. These parameters cannot be determined in the triaxial cell as sufficiently large displacements cannot be achieved to develop residual conditions.

For practical purposes failure envelopes are determined from a series of related tests. In the triaxial cell the most common tests are UU, CU and CD as detailed above. As far as possible three identical samples are obtained and tested at different cell pressures, giving three failure stress circles, and a failure envelope is drawn touching the three circles. Where three test specimens cannot be obtained, a multi-stage testing technique on one sample may be used (e.g. Kenney and Watson, 1961; Parry and Nadarajah, 1973). Failure envelopes for these different tests are discussed below.

2.3.1 UU tests

When an increment of isotropic external pressure is applied to a saturated soil, without permitting any drainage of water from the soil voids, this increment of pressure is carried entirely by the pore water because it is incompressible compared with the soil fabric. A fine-grained saturated soil specimen, taken from the field, when first set up on the triaxial pedestal has an internal negative pore pressure u_e (Figure 2.9(a)), the magnitude of which reflects the effective stresses in the ground (Section 4.4). The effective stress is therefore

$$\sigma'_a = \sigma'_r = -u_e$$

Under applied cell pressure, all the applied isotropic total stress σ_{cp} is taken by an increase in pore water pressure, and the effective stress remains equal to $-u_e$ (Figure 2.9). This is so, regardless of the magnitude of σ_{cp}, and as strength depends entirely on effective stress the measured strength is the same regardless of the applied cell pressure. Thus, if three identical specimens are tested under different cell pressures $\sigma_{cp}(1)$, $\sigma_{cp}(2)$, $\sigma_{cp}(3)$, three stress circles of equal radius will result, and an envelope to these will be a horizontal straight line, as shown in Figure 2.10. This is known as the $\phi_u = 0$ case, and the undrained shear strength is c_u. All three specimens are represented by the same effective stress Mohr circle, and the pore pressure at failure is the displacement of the relevant total stress circle relative to the effective stress circle. The value of $u_f(3)$ for applied $\sigma_{cp}(3)$ is indicated in Figure 2.10.

As there is only one effective stress circle for the three tests, an effective stress envelope cannot be drawn. However, for many soft clays c' is zero or very small in magnitude, and a line tangential to the effective stress circle passing through the origin can at least give a good indication of ϕ'. As a corollary to this, if the undrained shear strength c_u is measured in a triaxial test, and thus the total stress

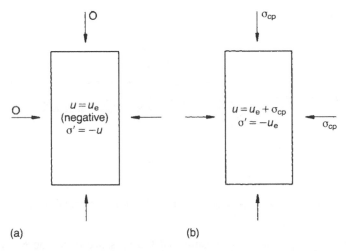

Figure 2.9 Pore pressure and effective stress in a triaxial test specimen of saturated clay: (a) immediately after extrusion from sampling tube; (b) after application of cell pressure σ_{cp}.

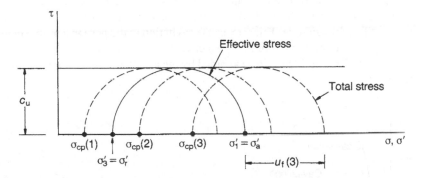

Figure 2.10 Stress circles at failure for identical saturated clay specimens tested under different triaxial cell pressures.

circle is known, the pore pressure at failure can be estimated if the effective stress envelope is already known. It is found by displacing the total stress circle along the σ axis until it just touches the envelope, as shown in Figure 2.11. In many clays it is possible to estimate ϕ' fairly closely and, assuming $c' = 0$, a reasonable estimate of u_f can be made, even if the precise effective stress envelope is not known.

In highly dilatant soils such as heavily overconsolidated clays or silts, negative pore pressure changes may occur in the test specimen under applied deviator stress, particularly at high shear strains. At high cell pressure this is not a problem as

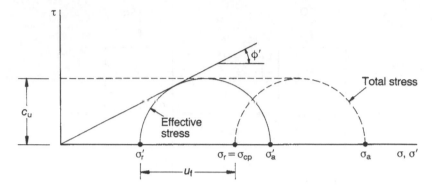

Figure 2.11 Determination of pore pressure from known ϕ' and total stress circle.

the initially high pressure will ensure that the pressure in the pore water will remain positive. It is usual, however, to perform one of the three tests, to establish the strength envelope, at a cell pressure corresponding to the total stress level in the ground, at the depth from which the specimen was recovered. In these circumstances the initial pore pressure after applying the cell pressure may be only a small positive value or even a negative value, so that negative pore pressure changes occurring under the applied deviator stress might result in negative pore pressures of large magnitude. This may cause cavitation in the pore water, in which case the $\phi_u = 0$ condition will not hold. An envelope such as that shown in Figure 2.12 will result. This problem can be avoided by the use of a suitable back pressure (Section 2.2).

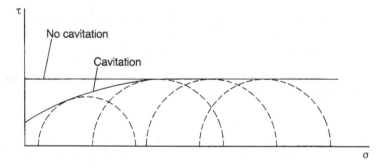

Figure 2.12 Effect of cavitation on undrained failure envelope.

EXAMPLE 2.2 STRENGTH PARAMETERS AND PORE PRESSURE IN UU TESTS

Three apparently identical specimens of undisturbed saturated clay are submitted to UU triaxial tests under cell pressures σ_{cp} of 50 kPa, 100 kPa and 200 kPa respectively. The axial stresses σ_{af} at failure are shown below.

Specimen	σ_{cp} (kPa)	σ_{af} (kPa)
1	50	238
2	100	320
3	200	396

Is it reasonable to assume this is a $\phi_u = 0$ case? If so, what is the value of c_u? Estimate the pore pressure at failure in each specimen if $c' = 0$, $\phi' = 24°$.

Solution

The stress circles at failure are plotted in Figure 2.13(a). Although a number of slightly different envelopes could be drawn, the variations in these would be 5° or less and it is logical to assume $\phi_u = 0$ and take an 'average' value of $c_u = 100$ kPa.

The pore pressure at failure for Specimen 1 can be found graphically, as shown in Figure 2.13(b), by finding the circle with radius equal to

$$\tfrac{1}{2}\left(\sigma_{af} - \sigma_{cp}\right) = 94 \text{ kPa}$$

with a centre on the $\tau = 0$ axis, which just touches the effective stress envelope. This is the effective stress circle for Specimen 1. The pore pressure at failure is then the horizontal distance between the centres of the total stress and effective stress circles. As the effective stresses exceed the total stresses, i.e. the effective stress circle is to the right of the total stress circle in Figure 2.13(b), the pore pressure is negative.

The location of the effective stress circle can also be found by simple trigonometry, as shown in Figure 2.13(c). It can be seen that the centre of the circle is given by

$$s' = 94 \operatorname{cosec} 24° = 231 \text{ kPa}$$

The centre of the total stress circle for Specimen 1 is given by

$$s = \tfrac{1}{2}(\sigma_{af} + \sigma_{cp}) = 144\,\text{kPa}$$
$$\therefore\ u_f = s - s' = -87\,\text{kPa}$$

The corresponding values for all three specimens are given below.

Specimen	s (kPa)	s' (kPa)	u_f (kPa)
1	144	231	−87
2	210	270	−60
3	298	241	+57

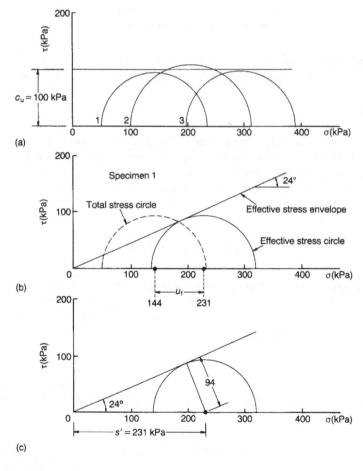

Figure 2.13 Example 2.2.

The negative pore pressure of 87 kPa in Specimen 1 is high, and the inability of the soil to maintain higher negative pore pressures than this may account for Specimen 1 exhibiting the lowest undrained strength of the three specimens.

2.3.2 CU tests

In consolidated undrained tests, three or more test specimens, if available, are initially submitted to different cell pressures, allowing dissipation of excess pore pressures to occur in each case. Each specimen is then subjected to an applied deviator stress in the manner of a UU test. If pore pressures are to be measured, a rate much slower than the normal rate for a UU test is necessary. The measurement of pore pressures greatly increases the value of the test, as it enables both total stress and effective stress strength parameters to be obtained.

In Figure 2.14, typical effective stress and total stress envelopes are shown. Figure 2.15 shows the corresponding plots of deviator stress and pore pressure against axial strain. It can be seen that the total stress envelope deviates markedly from linearity in the overconsolidated range, whereas the effective stress envelope remains almost linear, showing at most a slight deviation from the linear envelope through the origin, to give a small cohesion intercept c'. In the heavily overconsolidated region the two envelopes cross over, because pore pressure changes at failure, which are positive for normally consolidated (e.g. u_{f1} in Figure 2.15(a)) and lightly to moderately overconsolidated soil, become negative when the same soil is heavily overconsolidated (u_{f2} in Figure 2.15(b)).

Although the strength parameter ϕ_{cu} is not commonly used in design and analysis, the closely related parameter c_u/σ'_c can be very useful, for example in

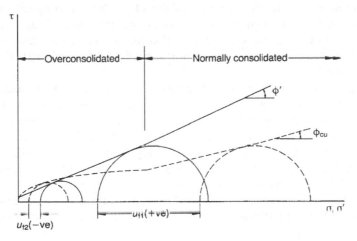

Figure 2.14 Typical total stress and effective stress envelopes for CU tests on clay specimens.

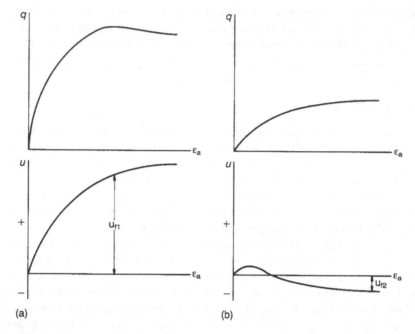

Figure 2.15 Typical plots of deviator stress and pore pressure against axial strain for undrained tests on: (a) normally consolidated clay; (b) heavily overconsolidated clay.

calculating the increase in shear strength in a soft clay under an embankment built up in stages, allowing consolidation to take place after each stage.

Another feature of interest to be seen in Figure 2.14 is the flatness of the total stress envelope for a lightly overconsolidated clay. This is readily explained in terms of the critical state concept, which predicts that under large shear strains, test specimens ultimately achieve a unique relationship between voids ratio and mean effective stress, and the corresponding shear strength is directly related to the mean effective stress.

In Figure 2.16, point A represents a normally consolidated specimen, points B and C lightly overconsolidated specimens and D, E moderately to heavily overconsolidated specimens. When tested in undrained shear, each specimen must follow a horizontal, constant e line until it reaches the critical state line (CSL). It will be seen that points A, B, C lie on a very flat part of the swelling curve and experience only slight differences in p' on reaching the CSL. Referring to equation 2.7, these specimens will thus show only slight differences in shear strength. As swelling continues, to points D and E in Figure 2.16, the curve steepens and, as expected, the total stress envelope in Figure 2.14 drops more rapidly as it approaches the origin.

The effective stress angle of shearing resistance ϕ' is widely used in design and analysis. As stated above it may deviate slightly from a straight line through the

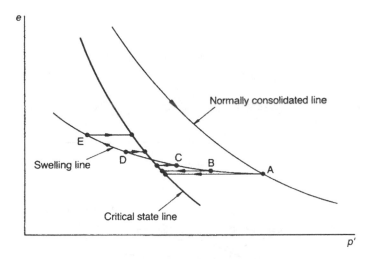

Figure 2.16 Critical state p'_{cs} values achieved in undrained tests.

origin for moderate to heavily overconsolidated soils, giving a small c' intercept. This can be significant in a low stress environment in the field, for example in the stability of shallow cuttings in heavily overconsolidated clay (Chandler and Skempton, 1974). In fact small c' values, which may be very significant in such field situations, are very difficult, if not impossible, to measure precisely in the laboratory, as they are obscured by secondary test effects such as friction in the test equipment, non-uniform strains and rate effects.

In Figure 2.17(a) total stress circles at failure are shown, with the resulting strength envelope, for a range of CU triaxial compression tests on remoulded Weald clay ($w_L = 46\%$, $w_P = 20\%$). All specimens were initially isotropically consolidated under cell pressures ranging from 20.7 kPa to 278 kPa, then different samples were allowed to swell to overconsolidation ratios ranging from 1 to 12 before performing the CU tests (Parry, 1956). The strong influence of OCR on the total stress envelope can be seen. A straight line touching the two circles for normally consolidated specimens passes through the origin, with $\phi_{cu} = 12.7°$. The corresponding value of $c_u/\sigma'_c = 0.28$.

The effective stress circles for the same test specimens are shown in Figure 2.17(b). It can be seen that the effective stress strength envelope is a straight line passing through the origin, despite the fact that OCR for the test specimens ranges up to 12. The effective stress angle of shearing resistance $\phi' = 23°$. Figure 2.17(b) displays the relationship between the total stress and effective stress envelopes, which intersect for samples with an OCR = 3.5. That is, specimens of this clay at OCR = 3.5, tested in undrained triaxial compression, show a net zero change in pore pressure between the start of the test and failure. Usually in this type of test the specimen will initially show some positive pore pressure change, but as the strain increases negative changes occur, as shown in Figure 2.15(b), and

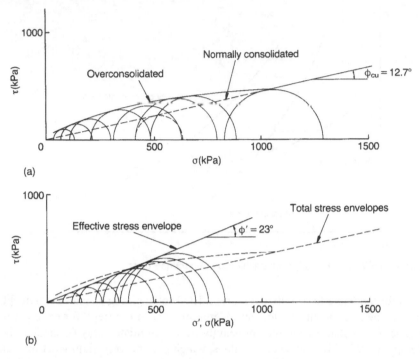

Figure 2.17 CU characteristics of remoulded Weald clay: (a) total stress; (b) effective stress.

thus, for an OCR = 3.5, these negative changes just balance the initial positive changes when failure is reached.

Most natural soft clay or silt deposits tend to exhibit some small degree of overconsolidation, despite the fact that they have not experienced any removal of overburden (Parry, 1970). This has some important practical implications, not least that small increases in effective stress in the field will not lead to an increase in undrained strength of the clay. This can be seen in Figure 2.18(a), where failure deviator stress is plotted against consolidation pressure for an undisturbed soft clay sample (w_L = 145%, w_p = 45%, w = 160%) from Launceston, Australia, taken from a depth of 2.6 m. The sample was submitted to a multi-stage CU test (Parry, 1968). Point 1 corresponds to the field moisture content and stress conditions. The initial flatness of the envelope for small increases in stress (explained above) means that where stage testing is being employed in the field to progressively strengthen the ground, the first metre or so of imposed loading will lead to little or no increase in strength.

Effective stress circles and the effective stress envelope for the same Launceston clay test are shown in Figure 2.18(b). A small c' intercept is seen, and, for a number of tests, ranged from 0 to 3.4 kPa, i.e. very small values lying within the range of possible test error. ϕ' in Figure 2.18(b) is 23° and over a number of tests ranged

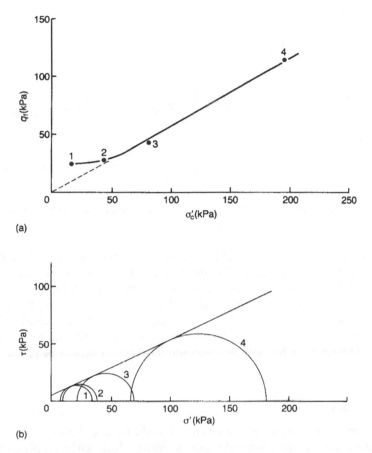

Figure 2.18 CU characteristics of undisturbed soft Launceston clay: (a) undrained shear strength vs consolidation pressure; (b) corresponding effective stress envelope.

from 21.5° to 25.5°. Back analysis of slips in low flood bank levees on this clay showed that good results were obtained from an effective stress analysis assuming $c' = 0$, $\phi' = 23°$. The pore pressures used in this analysis were measured directly by conducting full scale embankment loading tests in the field.

Data from CU tests on undisturbed specimens of a heavily overconsolidated clay are plotted in Figure 2.19 (Fugro, 1979). The test specimens were taken from a deposit of Gault clay near Cambridge, England, and the results shown in Figure 2.19 are from a sampling depth of 1.67 m ($w_L = 78\%$, $w_P = 27\%$, $w = 28\%$). This deposit is thought to have had a past cover of about 500 m of sediments.

The first sample of Gault clay was tested at the field voids ratio with initial pore pressure measured and no drainage allowed. The second and third samples were consolidated under higher cell pressures before conducting CU tests. The strong curvature of the total stress envelope is clearly seen. The effective stress

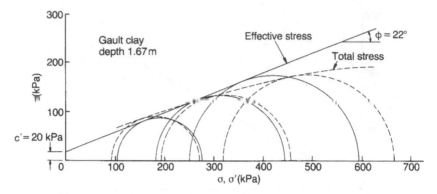

Figure 2.19 Total stress and effective stress CU characteristics of undisturbed stiff Gault clay.

envelope is linear with $\phi' = 22°$ and exhibits a c' intercept of 20 kPa. Back analysis of slips in nearby cuttings up to 3.2 m deep indicated field values of c' could be as low as 2 kPa (Parry, 1988), which is consistent with findings for other heavily overconsolidated clay deposits in South East England (Chandler and Skempton, 1974). The low values of c' obtained in the field are probably explained in large measure by the strong fissuring of these clays. It is clear that high laboratory measured values of c' for heavily overconsolidated clays cannot be relied upon in design.

2.3.3 CD tests

Drained triaxial tests are simpler to set up and conduct than CU tests; it is basically easier to measure volume change than pore pressure, which can be greatly affected by small amounts of air in the soil voids or the measuring system. On the other hand, drained tests take longer to perform and provide only effective stress parameters, in contrast to CU tests which provide information on both undrained strength parameters and effective stress strength parameters.

A typical envelope for CD tests on clay is shown in Figure 2.20 and corresponding plots of deviator stress and volumetric strain against axial strain are shown in Figure 2.21. For normally consolidated soil the envelope will usually be linear and pass through the origin. In the overconsolidated range the envelope will usually be essentially linear, but deviating from the normally consolidated envelope to give a c' intercept.

Stress circles and the corresponding strength envelope for a range of CD tests on isotropically normally consolidated and overconsolidated remoulded Weald clay (Parry, 1956) are shown in Figure 2.22. It can be seen that the normally consolidated specimens in Figure 2.22(a) give a linear envelope through the origin, with $\phi' = 21.2°$. This can be compared with the envelope for CU tests in Figure 2.17(b) giving $\phi' = 23°$ for the same clay. The deviation of the envelope

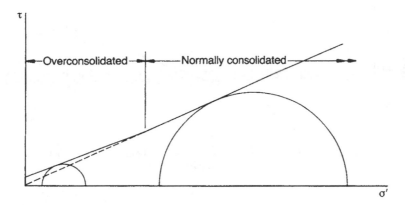

Figure 2.20 Typical failure envelope for CD tests on clay.

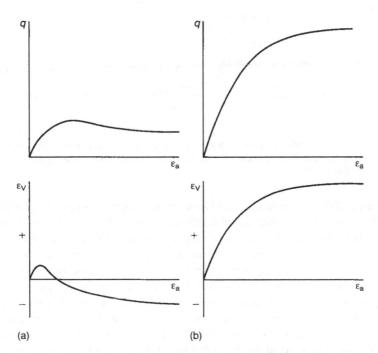

Figure 2.21 Typical plots of deviator stress and volumetric strain against axial strain for CD tests on: (a) heavily overconsolidated clay; (b) normally consolidated clay.

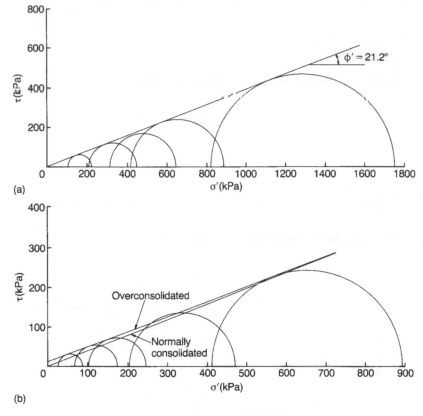

Figure 2.22 Failure circles and strength envelope for CD tests on remoulded Weald clay: (a) normally consolidated; (b) overconsolidated, showing deviation from (a).

in Figure 2.22(b) for overconsolidated specimens is very slight, giving $c' = 10$ kPa, which compares with indistinguishable envelopes for CU tests on normally consolidated and heavily overconsolidated specimens shown in Figure 2.17(b). The small cohesion intercept for the drained tests can be attributed entirely to dilatancy, which is suppressed in undrained tests. The circles shown for heavily overconsolidated specimens in Figure 2.22 were for peak failure conditions, when the specimens were continuing to show a volumetric increase. Further straining caused a progressive drop in deviator stress and decrease in rate of volume increase as the sample tended towards the critical state. As the heavily overconsolidated samples, under large axial strain, tended to deform unevenly, in some cases with the development of a distinct shear plane or zone, a true critical state could not be reached uniformly through the specimen. Nevertheless, at the end of the tests the stress circles lay very close to the normally consolidated envelope, which they would have just touched if the critical state had been reached.

The lower value of $\phi' = 21.2°$ in Figure 2.22(a) for the CD tests compared with $\phi' = 2.3°$ in Figure 2.17(b) for the CU tests may be explained, in part at least, by the slower rate of testing for the CD tests. The time to failure in CD tests ranged from 15 h to 100 h, compared to about 4 h for CU tests.

A series of CD tests on heavily overconsolidated undisturbed Gault clay (Clegg, 1981) gave ϕ' ranging from 22° to 24°, with c' consistently about 20 kPa. These results agreed closely with those from the CU tests (Figure 2.19), but it bears repeating that c' values of this magnitude are not realistic for use in design, at least for cuttings in strongly fissured clays of this type.

2.4 Triaxial extension tests

Although in some practical applications, such as investigating the soil strengths in the base of deep excavations, the triaxial extension test is more appropriate than triaxial compression, it is rarely performed in practice. It is a slightly more difficult test to perform than the compression test and, as the sample area is decreasing, it can exhibit some instability as failure is approached.

In triaxial extension:

$$\sigma_1 = \sigma_2 = \sigma_r \tag{2.13}$$

$$\sigma_3 = \sigma_a \tag{2.14}$$

In Figure 2.23 typical results which might be expected from UU, CU and CD extension tests (with decreasing axial stress) on a low to moderately sensitive, lightly overconsolidated, natural clay are plotted qualitatively, as deviator stress $(\sigma_r - \sigma_a)$ and pore pressure or volumetric strain against axial strain. Notable features are the development of strong negative pore pressures in UU and CU tests, leading to a plateau type failure, unless sample defects play a major part in the failure. In CD tests strong dilation occurs leading to a sharply peaked failure condition, which may be accompanied by necking of the failure specimen, or the development of a pronounced failure plane or failure zone. These characteristics become even more pronounced in heavily overconsolidated clays, and fissuring may also exert a major influence on the behaviour.

Assuming the critical state to hold, equations 2.7, 2.8 and 2.9 apply equally to extension tests as to compression tests, but in extension tests $\sigma_2' = \sigma_1'$, and equation 2.10 becomes

$$\left(\sigma_1' - \sigma_3'\right)_{cs} = \frac{M}{3}\left(2\sigma_1' + \sigma_3'\right)_{cs} \tag{2.15}$$

Substituting equation 2.15 into 2.11 gives

$$M = \frac{6 \sin \phi'}{3 + \sin \phi'} \qquad (2.16)$$

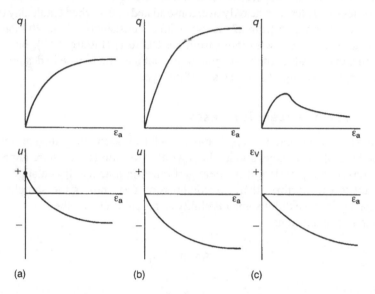

Figure 2.23 Typical plots for (a) UU, (b) CU and (c) CD tests on an undisturbed lightly overconsolidated clay.

Thus, on a q_{cs} vs p'_{cs} plot the critical state envelopes have different slopes in compression and extension, as shown in Figure 2.24(a). It has become conventional to plot compression above the $q = 0$ axis and extension tests below it, as shown in Figure 2.24(b), although of course there are positive and negative shear stresses acting within the specimens in both types of tests.

As undrained strengths c_u at the critical state are given by

$$c_u = \frac{q_{cs}}{2} = \frac{M}{2} p'_{cs} \qquad (2.17)$$

then from equations 2.12 and 2.16, identical soil specimens tested undrained in compression and extension will have different strengths in the ratio

$$\frac{c_u(\mathrm{E})}{c_u(\mathrm{C})} = \frac{3 - \sin \phi'_{cs}}{3 + \sin \phi'_{cs}} \qquad (2.18)$$

This assumes extension and compression values of ϕ'_{cs} to be the same. The ratios predicted by equation 2.18 for different ϕ'_{cs} values are given in Table 2.1.

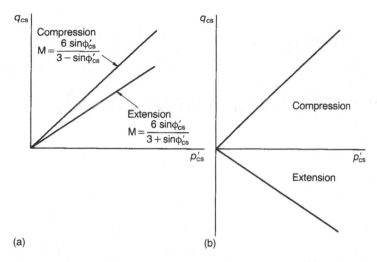

Figure 2.24 Comparison of triaxial compression and extension critical state strengths: (a) plotted in positive q_{cs} space; (b) conventional plot.

Table 2.1 Predicted ratios of undrained shear strengths

ϕ'_{cs}	$\dfrac{c_u(E)}{c_u(C)}$
20°	0.80
25°	0.75
30°	0.71
35°	0.68

Many workers have published relative values of undrained shear strengths in triaxial extension and compression for remoulded and natural clays, and for normally consolidated and overconsolidated specimens. These have invariably shown $c_u(E)$ to be less than $c_u(C)$ in the ratio ranging generally from 0.6 to 0.85. These include both peak and critical state values of c_u, so the predictions of relative undrained shear strengths made using the critical state theory apply also, in a general way at least, to peak conditions. In Figure 2.25 triaxial compression and extension undrained shear strengths are plotted against depth for heavily over-consolidated Gault clay for which $\phi' = 22°$ to 24°; and with the exception of one result at 4 m depth, extension strengths are 15% to 45% less than compression strengths, the strength ratio averaging about 0.7 (Fugro, 1979).

Published results present conflicting evidence regarding relative values of ϕ' in triaxial compression and extension tests on both clays and sands. In part this may be attributed to the instability which can develop in extension tests near

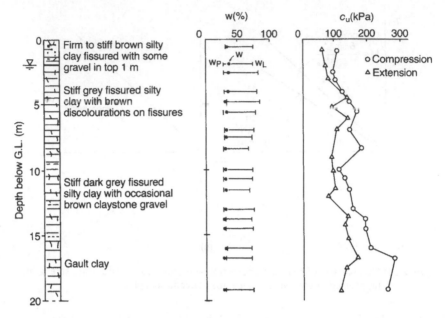

Figure 2.25 Profile of undrained triaxial compression and extension strengths for Gault clay.

failure, which tends to give a low value of $\phi'(E)$. In carefully controlled tests it is usual to find magnitudes of $\phi'(E)$ a few degrees higher than $\phi'(C)$, the difference generally increasing with increasing ϕ' (e.g. Parry, 1971; Reades and Green, 1976; Saada and Bianchini, 1975).

EXAMPLE 2.3 RELATIVE UNDRAINED STRENGTHS IN COMPRESSION AND EXTENSION

A saturated soil sample is known to have critical state ϕ' values of 30° in triaxial compression and 32.5° in triaxial extension. Assuming critical state conditions to hold, estimate the ratio of undrained strengths in triaxial compression and extension for initially identical test specimens.

Solution

$$M = \frac{6 \sin \phi'_{cs}}{3 - \sin \phi'_{cs}} \text{ in compression}$$

Equation 2.12:
$$= \frac{6 \sin 30°}{3 - \sin 30°}$$
$$= 1.20$$

Equation 2.16:
$$M = \frac{6 \sin \phi'_{cs}}{3 + \sin \phi'_{cs}} \text{ in extension}$$
$$= \frac{6 \sin 32.5°}{3 + \sin 32.5°}$$
$$= 0.91$$

Equation 2.17:
$$c_u(C) = 0.60 p'_{cs}$$
$$c_u(E) = 0.455 p'_{cs}$$

If the critical state holds, p'_{cs} is identical in compression and extension. Consequently:

$$\frac{c_u(E)}{c_u(C)} = \frac{0.455}{0.60} = 0.76$$

2.5 Influence of initial stress and structural anisotropy on strength of clays

Immediately following the deposition of a sedimentary soil deposit the vertical effective stress (usually assumed to be a principal stress) σ'_{v0} at any depth is greater than the horizontal effective stress σ'_{h0} at the same depth. The ratio

$$K_0 = \frac{\sigma'_{h0}}{\sigma'_{v0}} \tag{2.19}$$

is discussed in more detail in Section 4.2.2. Thus, an element of soil at any depth is subjected to an anisotropic stress and this, together with particle orientations taken up during the deposition process, produces structural anisotropy in the soil. It is difficult to separate stress and structural anisotropy completely, as any change in effective stress produces structural changes as well as stress changes. Subsequent removal of overburden causes the soil to become overconsolidated, leading to an increase in K_0 which, for heavily overconsolidated soil, can result in a horizontal effective stress much higher than the vertical effective stress. The overconsolidation process will change the particulate structure of the soil skeleton to some degree and may also lead to pronounced fissuring in clays. These fissures, which are usually fairly random in their orientation, can introduce marked weaknesses into the soil.

Triaxial test specimens cut at different inclinations from natural soft clay

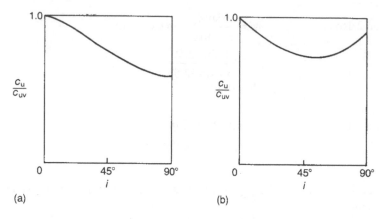

Figure 2.26 Examples of the influence of sample orientation on the measured undrained strengths of natural clays: (a) trend 1; (b) trend 2.

deposits, which are normally consolidated or lightly overconsolidated, usually show different undrained shear strengths in compression tests. Two different trends have been reported:

1. as shown in Figure 2.26(a), a progressive decrease in strength with increase in angle i between the vertical and the sample axis;
2. as shown in Figure 2.26(b), a decrease in strength from the vertical with increasing i, but reaching a minimum at $0 < i < 90°$, subsequently increasing in strength with further increase in i, but with the horizontal sample showing a smaller strength than the vertical sample.

These trends have been reported as shown in Table 2.2.

Trend (1) can be explained in terms of anisotropic stress path (Section 5.3), while trend (2) is probably due to particle orientations or thin horizontal weak

Table 2.2 Trends for triaxial test specimens cut at different inclinations from natural soft clay deposits.

Study	Soil	$\dfrac{c_{uh}}{c_{uv}}$	$\dfrac{c_{u\,min}}{c_{uv}}$
Trend 1			
Lo (1965)	Welland clay	0.77	–
De Lory and Lai (1971)	Welland clay	0.70	–
Wesley (1975)	Mucking Flats clay	0.64 to 0.78	–
Trend 2			
Duncan and Seed (1966)	San Francisco Bay mud	0.80	0.77
Parry and Nadarajah (1974b)	Fulford clay	0.84	0.70

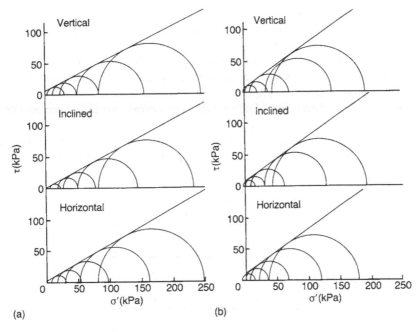

Figure 2.27 Influence of sample orientation and stress path on effective stress strength envelopes for soft Fulford clay: (a) compression; (b) extension.

layers in the clay lying close to the direction of potential failure planes in the soil.

There is conflicting evidence on the influence of initial stress and structural anisotropy on ϕ' at failure, but the bulk of currently available evidence suggests that anisotropy has very little influence on ϕ' measured in triaxial compression tests.

In Figure 2.27 stress circles and failure envelopes at failure for soft Fulford clay are shown (Parry and Nadarajah, 1974b). Samples taken vertically, inclined (45° to the vertical) and horizontally were submitted to consolidated undrained tests in triaxial compression and extension. Values of c', ϕ' are listed in Table 2.3.

Table 2.3 Values of c' and ϕ' from triaxial compression and extension tests

Sample orientation	Compression		Extension	
	c' (kPa)	ϕ'	c' (kPa)	ϕ'
Vertical	8	26°	9	34°
Inclined	3	27°	8	34°
Horizontal	4	29°	11	35°

The difference in ϕ' between compression and extension tests averages about 7°, but different sample orientations give a maximum difference of only 3° in compression and 1° in extension, showing the influence of initial anisotropy to be much less than the influence of stress direction in producing failure.

The differences in c' values may not be significant, as this parameter is very sensitive to slight changes in test data.

One noticeable effect of initial anisotropy is to increase the difference between ϕ' values in compression and extension, an effect which is well illustrated by the large difference of 7° in Table 2.2. This is again evident when comparing values of ϕ' for isotropically consolidated and K_0 consolidated, laboratory-prepared kaolin (Parry and Nadarajah, 1974a) as shown below.

	ϕ'	
	Compression	*Extension*
Isotropic	23°	21°
K_0	21°	28°

The lower value in extension than compression for isotropically consolidated specimens is unusual, but the most prominent feature is again the small spread of values for compression tests and the much higher $\phi'(E)$ for anisotropically consolidated specimens. Other workers have shown exactly the same trend for kaolin (Atkinson *et al.*, 1987).

2.6 Rupture planes in clays

For soils, the Mohr–Coloumb failure concept states that an element of soil will fail when the effective stress Mohr's circle for the element just touches the effective stress strength envelope. If the full stress circle is drawn it will touch the envelopes at two points with identical ratios of τ/σ'. The two planes through the element which then become rupture surfaces are those upon which this stress ratio τ/σ' acts.

This concept was put forward by Terzaghi (1936) and is shown in Figure 2.28 for triaxial specimens tested in compression and extension. Using the pole point method, it is easily shown that the directions of the rupture planes α, relative to the plane on which the principal effective stress σ_1' acts, are

$$\alpha\left(\mathrm{C}\right) = 45° + \frac{\phi'}{2} \tag{2.20a}$$

$$\alpha\left(\mathrm{E}\right) = 45° - \frac{\phi'}{2} \tag{2.20b}$$

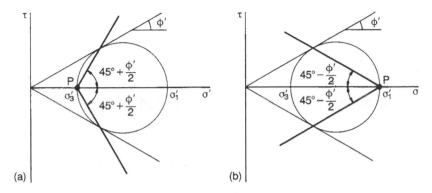

Figure 2.28 Orientations of rupture planes in: (a) triaxial compression tests; (b) triaxial extension tests. (Note: P is pole point for planes.)

It was argued by Terzaghi that the angle ϕ' in equations 2.20a and 2.20b should be the Hvorslev angle ϕ'_e, that is the angle of the effective stress envelope to circles with different stress conditions but identical water contents at failure. Gibson (1953) pointed out that ϕ'_e, is the sum of the resistance arising from frictional interaction between grains and a strength related to the rate of volume change of the test specimen at failure. He proposed the modified Hvorslev expression for shear box tests:

$$\tau_f = c_R + \sigma'_n \tan \phi'_R + \sigma'_n g_f \tag{2.21}$$

where $\sigma'_n g_f$ is the work expended at failure to expand the specimen against the applied stress σ'_n.

He determined ϕ'_R from drained shear box tests on a number of clays with ϕ'_R ranging from 3° to 26°, and also a sand mica with $\phi'_R = 30°$. Cylindrical specimens of each soil, 76 mm in height and 38 mm diameter, were prepared and submitted to unconfined axial compression at constant strain rate. Careful observations were made of the inclinations of failure planes as they developed. As the soils were saturated and the tests performed undrained, the inclinations of the failure planes were assumed to be reflecting the true friction angle (identical to ϕ'_e for undrained conditions). Good agreement was obtained, as shown in Figure 2.29, between ϕ'_R from the drained shear box tests and values deduced by applying equation 2.20a to observed inclinations of failure planes.

2.7 Shear bands

In a perfectly uniform, ideal cylindrical specimen of soil submitted to axial compression free of any restraints, slip should occur equally on all potential failure planes and consequently the specimen **should** deform as a perfect cylinder, free of discontinuities such as Mohr–Coulomb ruptures. This ideal deformation

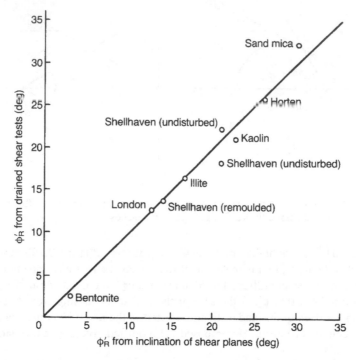

Figure 2.29 Comparison of friction angles from drained shear box tests with values deduced from inclinations of rupture planes. (After Gibson, 1953.)

geometry does not happen in the triaxial cell, even with the most carefully reconstituted laboratory specimens, because of boundary restraints imposed by the end platens and sample membrane. The specimens assume a barrel shape and in some cases, as failure is approached, distinct discontinuities develop, variously referred to by terms such as rupture surfaces, failure planes, failure zones or shear bands. These discontinuities, usually only one or two in number, initiate from points of weakness, which are inevitably present in natural undisturbed soils and even in specimens carefully constituted in the laboratory. In most cases the discontinuities are not simple surface-on-surface slips, but consist of a thin zone of soil, sometimes only a few grains in thickness, deforming to a much greater extent than the rest of the soil, and the term 'shear bands' has now become popularly applied to these. Many factors influence the pattern of shear bands, such as their number, thickness and orientation, if they do occur. These factors include particle size and size distribution, global uniformity or anisotropy, fissuring or other localized weaknesses, foreign inclusions in the soil, soil density, stress levels and stress history, the constraints imposed in loading the soil (in the laboratory or the field) and the rate of loading.

Although the Mohr–Coulomb concept of failure in soil is generally accepted, it does not necessarily follow that observed shear band orientations correspond to the orientations of planes of maximum effective stress obliquity. In some cases, at least, the directions correspond more closely with zero extension lines (see Section 1.13). In the direct shear box, the rupture discontinuities are constrained to follow the directions of zero direct strain. Skempton (1967) envisioned, for clays, five stages in the development of the shear band in the direct shear box, the essential features of which are the development of (stress related) Reidel shears at or near peak strength, inclined to the direction of imposed shear movement, followed by the development of 'displacement shears' in the direction of movement, as further slip along Reidel shears becomes kinematically impossible. Skempton concludes that the shear band for clay in the shear box is typically 10 to 50 μm wide and, after sufficient movement, particles within the band become strongly oriented and the strength of the specimen falls to its residual value.

In the triaxial cell the soil specimen has greater freedom to assume a preferred slip direction than in the direct shear box, although the end platens and membrane still impose some boundary restraints. The direction of zero extension is 55° to the horizontal for an undrained compression test, assuming uniform deformation of the test specimen. In fact, water migrations can take place within a globally undrained specimen, particularly at slow rates of strain, causing local differences in the angle of dilation, which can influence the orientations of shear bands as demonstrated by Atkinson and Richardson (1987). Other influences on shear band orientation include boundary restraints, barrelling prior to rupturing and imperfections in test specimens. As the angles of the Mohr–Coulomb planes given by equation 2.20a are 55° and 60° for friction angles of 20° and 30° respectively, it is clearly difficult to determine if observed shear bands are stress generated or strain generated. It is also conceivable that initial minor stress related slips, barely detectable or not detectable at all, merge into a clearly observed displacement related rupture and the slight waviness or curvature often seen in shear bands may give credence to this. It may also reconcile the observations by Gibson discussed above with the conclusions of Atkinson and Richardson (1987) and Atkinson (2000) that shear band orientations are governed by the angle of dilation. The difficulty of determining if shear bands are stress or strain generated is complicated by the fact that, in the triaxial test, mobilized friction on planes with inclinations ± 10° to the plane of maximum obliquity is within 0.5° of the full friction angle. Thus, very minor anisotropies or imperfections in a test specimen have the potential to influence the inclination of shear bands, if stress generated, by several degrees.

It has been shown by Atkinson (2000) that the rate of strain applied to an unconfined specimen of overconsolidated kaolin can determine whether or not a shear band will form. In the example cited a slow rate of strain, allowing local dilation to take place, resulted in a very distinct rupture discontinuity, whereas a fast strain rate produced only barrelling. In normally or lightly overconsolidated specimens the opposite behaviour might be expected, as slow strain rates would allow migration of water away from, and thus strengthening of, weak

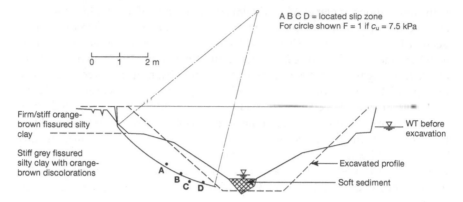

Figure 2.30 Investigation of shallow slip in heavily overconsolidated Gault clay (Parry, 1988).

zones. At normal rates of strain employed in the laboratory, normally or lightly overconsolidated laboratory prepared specimens usually exhibit barrelling, as do, quite commonly, natural undisturbed specimens, although these are more likely to have defects or anisotropies leading to the formation of rupture discontinuities or shear bands. Burland (1990) cites an interesting set of experiments showing that laboratory prepared, anisotropically normally consolidated specimens of kaolin, submitted to undrained triaxial tests, were more likely to develop shear bands than corresponding isotropically prepared specimens.

The formation of shear bands in clay deposits in the field can be much more dramatic than in the laboratory, because of the greater volume of soil involved and the possibility of greater and slower shear movements. Henkel (1956) measured water contents in a thin shear band behind a failed retaining wall 10% higher than the values away from the shear zone. The present author (Parry, 1988) has found even more pronounced dilatancy and softening in a study of shallow slips in channels excavated into heavily overconsolidated, highly fissured Gault clay at a site near Cambridge. A typical slip profile is shown in Figure 2.30. Inspections and measurements of the shear band were made in trenches at points A, B, C and D a few months after the most recent movements, which occurred progressively and intermittently after periods of heavy rain over a period of some two years before the site investigation. It proved a simple matter to locate the shear band, found to be about 80 mm thick, by means of a sharpened surveyor's stake, which could easily be pushed by hand into the softened soil in the sheared zone, but could not penetrate the undisturbed soil below it or the partially disturbed soil above it. The soil in the sheared zone consisted of a matrix of very soft clay with small, harder pieces of clay, which had broken away on fissures. Moisture contents w were determined within the shear band and at points 150 mm above it and 150 mm below it, together with small field vane strengths c_{uv}. Typical values were:

	w (%)	c_{uv} *(kPa)*
150 mm above shear band	32–38	50–75
Within shear band	43–48	9–14
150 mm below shear band	33–35	120

The high moisture content in the shear band attests to strong dilatancy occurring during its formation, leading to an order of magnitude reduction in undrained shear strength. Back analysis of the slips gave a value of undrained shear strength equal to 7.5 kPa for the original profile and 6.0 kPa for the final profile. This suggests some strengthening of the soil in the shear band took place in the few months between the cessation of slipping and conducting the site investigation.

Roscoe (1970) has described model tests in the laboratory, pushing a 330 mm high wall into sand with lead shot embedded in it, allowing shear strains and volumetric strains to be obtained from radiographs. Tests were performed rotating the wall about its top and about its toe and translating the wall into the sand. Contour plots of shear strain and volumetric strain showed very clearly the development of shear bands and they could also be seen as dark bands on the radiographs themselves. Roscoe concluded from this work that the rupture surfaces coincided with directions of zero extension and that, in any element within the soil mass, rupture would occur in a direction of zero extension at the instant the Mohr–Coulomb failure condition was reached on any plane within the element.

The experimental observation by Roscoe (1970) that shear bands in sands are about 10 grains thick has been supported by Mühlhaus and Vardoulakis (1987). They observed thickness of about eight grains, which was not affected by any geometrical dimensions of the soil body. In a series of plane strain tests on loose sand, mostly undrained, Finno *et al.* (1997) observed shear bands of width 10 to 25 mean grain diameters inclined at angles ranging from 55° to 65° to the direction of minor principal stress, and thus corresponding more closely to the directions of Mohr–Coulomb rupture than to zero extension directions of 45° to the direction of minor principal stress if the specimens were deforming uniformly throughout. However, a close examination of strain components within the bands indicated essentially zero normal strains parallel to the bands, leading Finno *et al.* to conclude that bands formed along lines of zero extension.

A programme of triaxial tests on sand with varying test conditions (Desrues *et al.*, 1996) demonstrated the complex and uncertain nature of shear band development. Tests were conducted on dense and loose sands, different length specimens, lubricated and non lubricated ends, homogeneous specimens and a specimen with a small foreign axial inclusion. It was observed that conditions favouring symmetry, such as end-platen friction and short specimens, exhibited multiple shear banding but were bad practice, inducing artificially high strengths.

Conditions upsetting symmetry, such as uneven densities or weaknesses in the specimen, bad centring or tilting of platens, produced single pronounced shear bands.

2.8 Influence of dilatancy on ϕ' for sands

Observing that the measured point of peak shear strength in sand usually coincided with the maximum rate of dilation in the test specimen, Bolton (1986) examined the concept of the strength parameter ϕ' consisting of a basic critical state value ϕ'_{cs}, modified by the maximum dilatancy angle ψ_{max} (equation 1.39). Putting

$$\phi' = \phi'_{cs} + 0.8\psi_{max} \tag{2.22}$$

he found that this relationship gave results virtually coincident with those given by the stress dilatancy theory of Rowe (1962).

Recognizing that ψ was not generally known, and that the dilatancy rate depended upon relative density of the sand I_D and the mean effective stress at failure p'_f, Bolton analysed results from 17 different sands tested in 12 different laboratories to establish a relationship between $(\phi' - \phi'_{cs})$, I_D and p'. As a result of this analysis he proposed the use of a dilatancy index I_R where

$$I_R = I_D(10 - \ln p') - 1 \tag{2.23}$$

The relative density I_D is given by

$$I_D = \frac{e_{max} - e}{e_{max} - e_{min}} \tag{2.24}$$

where e is the voids ratio. The dimensions of p' in equation 2.23 are kPa and the constants 10 and -1 were fixed by establishing the best fit with published experimental results.

The experimental data yielded the following empirical best fit relationships (Figure 2.31).

1. Plane strain:

$$\phi' - \phi'_{cs} = 5I_R \quad \left(\text{in degrees}\right) \tag{2.25}$$

2. Triaxial compression:

$$\phi' - \phi'_{cs} = 3I_R \quad \left(\text{in degrees}\right) \tag{2.26}$$

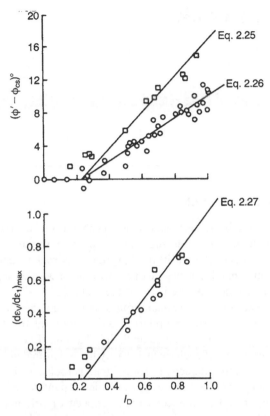

Figure 2.31 Influence of relative density of sand on friction angle and dilatancy rate. (After Bolton, 1986.)

3. All conditions:

$$-\left(\frac{\delta\varepsilon_V}{\delta\varepsilon_1}\right)_{max} = 0.3 I_R \qquad (2.27)$$

In order to determine ϕ' using these equations it is necessary to know the magnitude of the critical state ϕ'_{cs}. From his study of published experimental data, Bolton proposed that in the absence of experimental information it is reasonable to take $\phi'_{cs} = 33°$ for a predominantly quartz sand and $\phi'_{cs} = 40°$ for a predominantly felspar sand.

For very loose sands, or sands under extremely high pressure, I_R may be negative, implying decreasing volume under shear, tending towards zero volume change as critical state is approached. Values of I_R greater than 4 imply extreme dilatancy and should be discounted, unless supporting data are available.

Chapter 3

Failure in rock

3.1 The nature of rock

The term 'rock' as used by geotechnical engineers covers a wide range of materials, from hard polyminerallic igneous and metamorphic rocks, such as granite and dolerite, on the one hand, to soft lithified argillaceous rocks, such as mudstone and shale, on the other hand. The degree of oxidization or weathering can also have a marked influence on the characteristics of the rock.

Intact unweathered igneous or metamorphic rocks are generally characterized by close contact along crystal boundaries and contain, at most, unconnected macropores which have little influence on their behaviour. In the mass, however, the engineering behaviour of such rocks is likely to be dominated by the presence of joints, fissures and other defects. In some igneous rocks, such as basalt which has flowed into water, vesicles may be present which can influence their behaviour.

Arenaceous rocks and argillaceous rocks consist of primary rock-forming minerals or secondary clay minerals, cemented to a greater or lesser degree at their contacts, but otherwise separated by interconnected voids. The fluid in the voids influences the behaviour of the intact rock, as it does in soil.

3.2 Intrinsic strength curve

Rupture of intact cylindrical rock specimens may be effected by uniaxial tension, unconfined axial compression, or confined compression such as in the triaxial cell. Typical examples of effective stress circles at rupture for the three types of tests are illustrated in Figure 3.1. Jaeger (1972) has termed the envelope to the stress circles at rupture the 'intrinsic curve'. In most cases it is found to approximate to a parabolic shape.

3.3 Griffith crack theory

It was postulated by Griffith (1921, 1924) that in brittle materials such as glass, fracture initiated when the tensile strength was exceeded by stresses generated at the ends of microscopic flaws in the material. These microscopic defects in intact

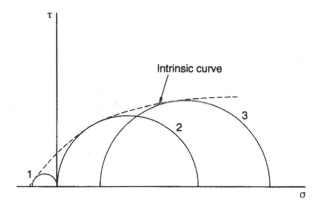

Figure 3.1 Typical rupture circles for rock and the Jaeger 'intrinsic curve': 1, uniaxial tension; 2, uniaxial compression; 3, triaxial compression.

rock could be small cracks, fissures or grain boundaries. As discussed by Hoek (1968), the theory predicts a parabolic failure envelope and thus yields an envelope of the same general form as that observed for intact rocks.

Griffith originally applied his theory to a plate of uniform thickness subjected to a uniaxial tensile stress, having an elliptical crack within the plate at right angles to the direction of loading. Subsequently he extended the theory to the extension of a crack within a plate subjected to compressive stresses σ_1, σ_2 (Figure 3.2). Assuming plane compression, he obtained the expression (Griffith, 1924)

$$\left(\sigma_1 - \sigma_2\right)^2 - 8T_0\left(\sigma_1 + \sigma_2\right) = 0 \tag{3.1}$$

where T_0 is the uniaxial tensile strength of the uncracked material (expressed as a positive number). This gives the strength envelope shown in Figure 3.3(a).

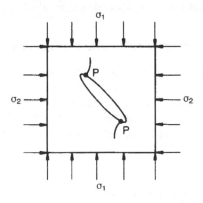

Figure 3.2 Griffith elliptical crack. (P indicates point of failure.)

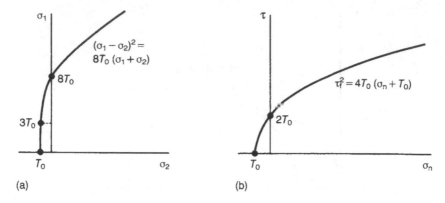

Figure 3.3 Griffith's strength envelopes for plane compression in terms of: (a) principal stresses σ_1, σ_2; (b) shear and normal stresses, τ, σ_n.

It will be seen that this theory predicts a uniaxial compressive stress at crack extension equal to eight times the uncracked uniaxial tensile strength.

Expressed in terms of the shear strength τ_f and the normal stress σ_n acting on the plane containing the major axis of the crack, the criterion becomes

$$\tau_f^2 = 4T_0\left(\sigma_n + T_0\right) \tag{3.2}$$

giving the strength envelope shown in Figure 3.3(b).

Although the Griffith criterion gives a strength envelope of the general form observed for tests on rock specimens, it does not provide a good model, in part because it ignores frictional forces on closed cracks. Modifications to the Griffith criterion taking account of friction on closed cracks and the influence of the intermediate principal stress (see presentations in Jaeger and Cook, 1976) have not led to satisfactory agreement with experimental results.

3.4 Empirical strength criteria for rock masses

As attempts to match a wide range of rock strength data with modified expressions based on the Griffith crack theory have not been successful, workers in the field have resorted to empirical expressions using the Griffith theory simply as a conceptual starting point. Hoek and Brown (1980) found that the peak triaxial strengths of a wide range of rock materials could be reasonably represented by the expression

$$\sigma_1' = \sigma_3' + \left(m\sigma_c\sigma_3' + S\sigma_c^2\right)^{1/2} \tag{3.3}$$

where σ_c is the unconfined uniaxial compressive strength of the intact rock making up the rock mass. This expression can also be presented in a normalized form by putting $\sigma'_{1N} = \sigma'_1/\sigma_c$, $\sigma'_{3N} = \sigma'_3/\sigma_c$ thus

$$\sigma'_{1N} = \sigma'_{3N} + \left(m\sigma'_{3N} + S\right)^{1/2} \tag{3.4}$$

The value of m ranges from about 0.001 for highly disturbed rock masses to about 25 for hard intact rock. The value of S ranges from 0 for jointed rock masses to 1.0 for intact rock.

The special cases of unconfined compressive strength σ_{cm} and uniaxial tensile strength σ_{tm} for a rock mass can be found by putting $\sigma'_3 = 0$ and $\sigma'_1 = 0$ respectively into equation 3.3. Thus

$$\sigma_{cm} = \left(S\sigma_c^2\right)^{1/2} \tag{3.5}$$

$$\sigma_{tm} = \tfrac{1}{2}\sigma_c\left[m - \left(m^2 + 4S\right)^{1/2}\right] \tag{3.6}$$

The general case of equation 3.4 and the special cases given by equation 3.5 and equation 3.6 are depicted in Figure 3.4.

Although equation 3.4, expressed in terms of principal effective stress, is useful in designing for confined situations in rock, such as underground excavations, it is not a convenient form for designing rock slopes, where the shear strength of a failure surface under a specified effective normal stress is required. A more convenient form of equation 3.4, which can be expressed in the form of a Mohr failure envelope, has been derived by Bray (Hoek, 1983):

$$\tau_{fN} = \left(\cot \phi'_i - \cos \phi'_i\right)\frac{m}{8} \tag{3.7}$$

where τ_{fN} is the normalized shear stress at failure ($= \tau_f/\sigma_c$) and ϕ'_i, is the instantaneous friction angle given, as shown in Figure 3.5, by the inclination of the tangent to the Mohr failure envelope for specific values of τ_{fN} and the effective normalized normal stress σ'_{nN} ($= \sigma'_n/\sigma_c$).

The instantaneous friction angle ϕ' is given by

$$\phi'_i = \tan^{-1}\left[4h\cos^2\left(30° + \tfrac{1}{3} \sin^{-1} h^{-3/2}\right) - 1\right]^{-1/2} \tag{3.8}$$

where

$$h = 1 + \frac{16\left(m\sigma'_{nN} + S\right)}{3m^2} \qquad (3.9)$$

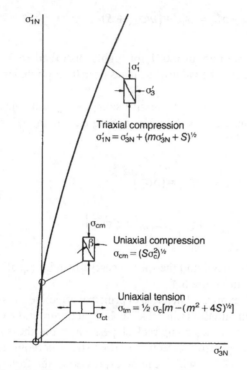

Triaxial compression
$\sigma'_{1N} = \sigma'_{3N} + (m\sigma'_{3N} + S)^{\frac{1}{2}}$

Uniaxial compression
$\sigma_{cm} = (S\sigma_c^2)^{\frac{1}{2}}$

Uniaxial tension
$\sigma_{tm} = \frac{1}{2}\sigma_c[m - (m^2 + 4S)^{\frac{1}{2}}]$

Figure 3.4 Hoek and Brown strength envelope. (After Hoek, 1983.)

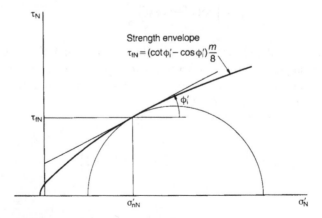

Strength envelope
$\tau_{fN} = (\cot\phi'_i - \cos\phi'_i)\dfrac{m}{8}$

Figure 3.5 Bray strength envelope. (After Hoek, 1983.)

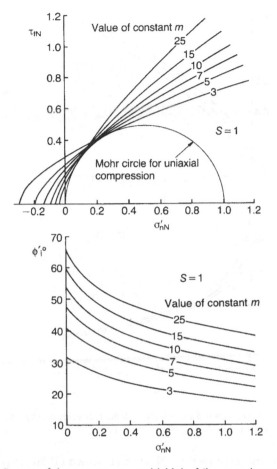

Figure 3.6 Influence of the constant *m* on: (a) Mohr failure envelopes; (b) instantaneous friction angle ϕ'_i, for s =1. (After Hoek, 1983.)

The influences of constants *m* and *S* on the Mohr failure envelopes and on the instantaneous friction angle are shown in Figure 3.6 and Figure 3.7 respectively. Hoek (1983) very approximately likens *m* to the angle of friction ϕ' and *S* to the cohesive strength intercept *c'*.

3.5 Empirical strength criteria for intact rock

For intact rock $S = 1$, and equation 3.4 becomes

$$\sigma'_{1N} = \sigma'_{3N} + \left(m\sigma'_{3N} + 1\right)^{1/2} \qquad (3.10)$$

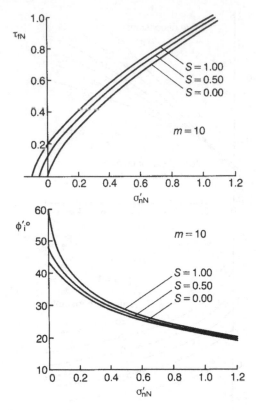

Figure 3.7 Influence of the constant S on: (a) Mohr failure envelopes; (b) instantaneous friction angle ϕ_i', for $m = 10$. (After Hoek, 1983.)

Published strength data examined by Hoek and Brown (1980) suggested that m increased with rock quality. Representative values for different rock groups are given below.

- Group a $m \approx 7$ for carbonate rocks with well developed crystal cleavage (dolomite, limestone, marble);
- Group b $m \approx 10$ for lithified argillaceous rocks (mudstone, siltstone, shale, slate);
- Group c $m \approx 15$ for arenaceous rocks with strong crystals and poorly developed crystal cleavage (sandstone, quartzite);
- Group d $m \approx 17$ for fine-grained polyminerallic igneous crystalline rocks (andesite, dolerite, diabase, rhyolite);
- Group e $m \approx 25$ for coarse-grained polyminerallic igneous and metamorphic rocks (amphibolite, gabbro, gneiss, granite, norite, quartz-diorite).

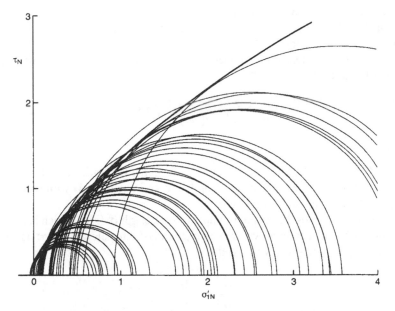

Figure 3.8 Mohr failure circles for five granites. (After Hoek, 1983.)

In Figure 3.8 Mohr stress circles at the peak failure state are shown for five different intact granites tested in the USA and the UK. They show excellent consistency, well represented by a failure envelope given by equations 3.7, 3.8 and 3.9, with substituted values $\sigma_c = 1$, $m = 29.2$, $S = 1$, giving a correlation of 0.99.

A corresponding plot for 11 different limestones given in Figure 3.9 shows much less consistency, which Hoek (1983) attributes to the fact that the generic term 'limestone' covers a range of carbonate rocks, with both organic and inorganic

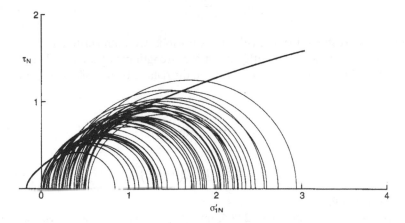

Figure 3.9 Mohr failure circles for 11 limestones. (After Hoek, 1983.)

origins and varying mineral composition, grain size and shape, and nature of cementation. Values of m derived from the test data ranged from 3.2 to 14.1 with a correlation coefficient of 0.68.

An alternative expression for intact rock has been presented by Johnston and Chiu (1984) and Johnston (1985):

$$\sigma'_{1N} = \left(\frac{M}{B} \sigma'_{3N} + 1 \right)^R$$

(3.11)

where M and B are the rock 'constants', the general trends for which are a decreasing B with increasing shear strength and an increasing M with increasing strength. M also depends on rock type.

Putting $\sigma'_{3N} = 0$ gives the uniaxial compressive strength $\sigma_c = \sigma'_1$, which is correct. The ratio of uniaxial compressive strength σ_c to uniaxial tensile strength σ_t is found by putting $\sigma'_1 = 0$, $\sigma'_3 = \sigma_t$, which yields

$$\frac{\sigma_c}{\sigma_t} = -\frac{M}{B}$$

(3.12)

An advantage of equation 3.11 over equation 3.10 is that it can be fitted to a wide range of strength envelope shapes by changing the power index B. Putting $B = 1$ gives a linear envelope:

$$\sigma'_{1N} = M\sigma'_{3N} + 1$$

(3.13)

which for

$$M = \frac{1 + \sin \phi'}{1 - \sin \phi'}$$

(3.14)

corresponds exactly to the normalized linear Mohr–Coulomb criterion.

By fitting equation 3.11 to effective stress strength data from a wide range of clay soils and rocks, Johnston (1985) found the data to be reasonably represented by

$$B = 1 - 0.0172 \left(\log \sigma_c \right)^2$$

(3.15)

$$M = 2.065 + 0.276 \left(\log \sigma_c \right)^2$$

(3.16)

where σ_c is in kPa.

As $\sigma_c \to 0$, $B \to 1$ and $M \to 2.065$, implying a virtually linear envelope with $\phi' = 20°$ for soft normally consolidated clay. In stiff heavily overconsolidated clays, typically $\sigma_c = 200$ kPa, giving $B = 0.9$ and thus a slightly curved envelope; while in hard rocks, typically $\sigma_c = 250$ MPa and $B = 0.5$, which is the familiar parabolic form.

Equation 3.16 was derived by assuming a parabolic relationship between M and $\log \sigma_c$, then fitting this to a wide scatter of data. It was found by Johnston (1985) to vary with rock type. For the same rock groupings as categorized by Hoek (1983) and listed above, the best-fit parabolic curves were found by Johnston to be:

- Group a $$M = 2.065 + 0.170\left(\log \sigma_c\right)^2 \tag{3.17a}$$

- Group b $$M = 2.065 + 0.231\left(\log \sigma_c\right)^2 \tag{3.17b}$$

- Group c $$M = 2.065 + 0.270\left(\log \sigma_c\right)^2 \tag{3.17c}$$

- Group e $$M = 2.065 + 0.659\left(\log \sigma_c\right)^2 \tag{3.17e}$$

Figure 3.10 shows published Mohr stress circles at peak failure for four different rock types: limestone, mudstone, sandstone and granite. In each case the solid line

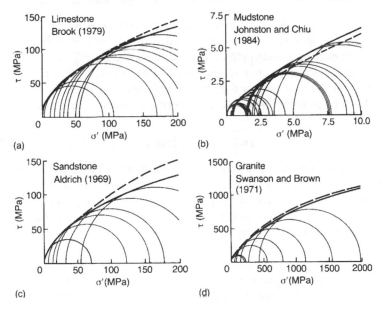

Figure 3.10 Failure circles for four different rocks with envelopes for Johnston 'best-fit' (solid line) and 'generalized' (broken line) B and M values. (After Johnston, 1985.)

Table 3.1 Best-fit and general parameters for four rock types

Rock type	Group	Best-fit σ_c (MPa)	Best-fit parameters		General parameters	
			B	M	B (eq. 3.15)	M (eq. 3.17)
Limestone	a	96	0.481	7.43	0.573	6.29
Mudstone	b	1.3	0.750	6.16	0.833	4.30
Sandstone	c	68	0.444	11.4	0.598	8.37
Granite	e	230	0.538	15.6	0.505	21.02

envelope shown is that given by equation 3.11, using best-fit B and M values, while the broken line envelope is that given by equation 3.11 using the generalized B and M values from equations 3.15 to 3.17.

This criterion predicts the generally observed variations in uniaxial compressive and tensile strength and also links soils, soft rocks and hard rocks. The actual parameters adopted are given in Table 3.1.

3.6 Strength of rock joints

3.6.1 Infilled joints

If the joint within a rock mass is infilled with a material weaker than the host rock then, depending upon the orientation of applied stress, the joint is likely to have a strong influence on the strength of the rock mass. Such infill material could be of a granular or cohesive nature, washed into the joint, or formed *in situ* by weathering of the host rock at the joint boundaries. The shear strength of this material will be governed by the failure criterion for soils,

$$s = c' + \left(\sigma_n - u\right)\tan\phi' \tag{3.18}$$

where c', ϕ' are the effective stress cohesion and friction parameters, σ_n is the total normal stress acting across the joint and u is the water pressure within the joint.

If the joint is infilled with clay, then for rapid transient loading, or if confinement of the clay restricts or prevents drainage during shear, it may be appropriate to adopt the undrained shear strength.

It has been shown in laboratory studies (Ladanyi and Archambault, 1977) that the shear strength of an infilled discontinuity will usually be higher than that of the filling material, as some contacts will occur between rock surface asperities. It follows, too, that the joint strength decreases with increased thickness of joint filling.

Much of the research on the shear strength of filled rock joints has concentrated on the influence of joint wall roughness, expressed as asperity height a, compared

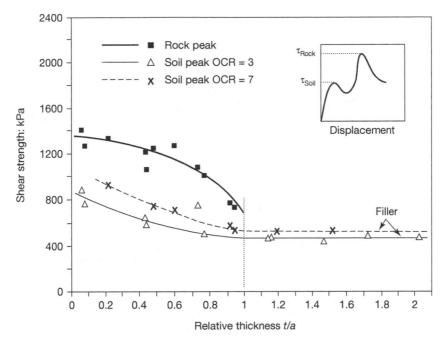

Figure 3.11 Strength of clay infilled sandstone joint tested in a ring shear device under normal pressure 1000 kPa (Toledo and de Freitas, 1993). Reproduced by kind permission of Thomas Telford.

to the thickness t of the fill material. An example of the influence of the relative thickness t/a for a series of tests on clay infilled sandstone joints is shown in Figure 3.11 (Toledo and de Freitas, 1993), and Figure 3.12 shows a strength model presented by the same authors embodying in a general way their own observations and those of a number of other researchers, embracing mica filler and plaster as rock (Goodman, 1970), kaolin clay and concrete blocks (Ladanyi and Archambault, 1977), kaolin and hard gypsum rock (Lama, 1978), oven dried bentonite and gypsum rock (Phien-Wej *et al.*, 1990) and fillers of kaolin, marble dust and fuel ash with plaster and cement as rock simulating joint materials.

The inset diagram in Figure 3.11 shows the infill soil providing the initial resistance to joint displacement and reaching a peak τ_{Soil} at small displacement, followed by a decline in strength depending on the brittleness of the infill soil, then an increase in strength with increasing displacement brought about initially by a flow of infill material from highly stressed to low stressed zones leading to contact of the rock asperities and a second, higher, peak strength τ_{Rock} depending on the rock properties. The decline in joint strength with increasing t/a is seen in Figure 3.11 up to about $t/a = 1$, after which the strength remains essentially constant. It is noticeable that the peak rock strength still exceeds that of the infill at $t/a = 1$. These features are incorporated into Figure 3.12, but it is noticeable that in this generalized

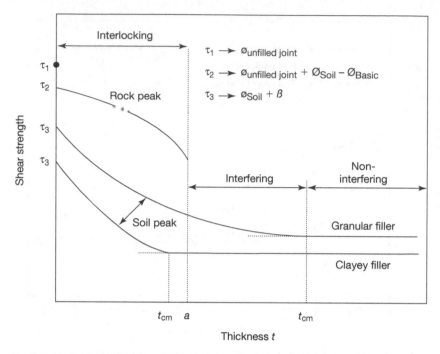

Figure 3.12 Generalized strength model for infilled joint (Toledo and de Freitas, 1993). Reproduced by kind permission of Thomas Telford.

model the critical thickness at which the soil peak reaches its minimum may be less than the asperity height for clayey infill and greater than asperity height for granular infill. A further feature of Figure 3.12 is the division of the different thicknesses into three ranges as proposed by Nieto (1974), namely interlocking in which the rock surfaces come into contact, interfering when there is no rock contact but the joint strength exceeds that of the infilling, and non-interfering when the joint strength equals that of the filler.

3.6.2 Clean joints

A simple expression for shear strength along a clean discontinuity in rock has been suggested by Patton (1966):

$$\tau_{jf} = \sigma_n' \tan\left(\phi_b' + i\right) \tag{3.19}$$

where σ_n' is the effective normal stress on the discontinuity, σ_b' is the 'basic' friction angle between smooth, dry surfaces of the rock and i is an equivalent angle representing the roughness of the surfaces. In effect i is a dilation angle, as shown in the idealized surface roughness models in Figure 3.13.

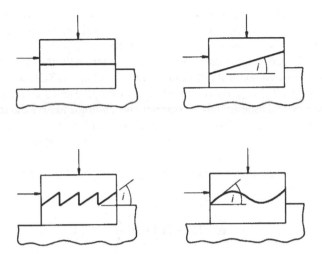

Figure 3.13 Dilation angle *i* for idealized surface roughness models. (After Brady and Brown, 1992).

As the normal effective stress increases, it eventually reaches a critical value, above which interaction of asperities forces the shear failure to occur partly through the asperities themselves, giving the bilinear plot of shear stress against normal effective stress shown in Figure 3.14. Overlapping of the two mechanisms, which will normally occur, leads to a curved relationship rather than a bilinear one.

An empirical expression for peak shear strength of saturated joints based on equation 3.19 has been suggested by Barton (1973):

$$\frac{\tau_{jf}}{\sigma'_n} = \tan\left[(JRC)\log_{10}\left(\frac{JCS}{\sigma'_n}\right) + \phi'_r\right] \qquad (3.20)$$

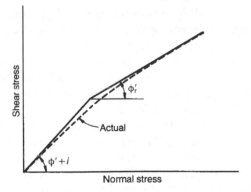

Figure 3.14 Bilinear model and actual clean joint strength.

where τ_{jf} is the shear strength; σ'_n the normal effective stress; JRC is the joint roughness coefficient; JCS is the effective joint wall compressive strength (saturated); ϕ'_r is the residual friction angle (wet, residual, drained).

Equation 3.20 incorporates three components of strength, the basic frictional angle ϕ'_b, a geometric component reflecting surface roughness (JRC) and an asperity failure component (JCS/σ'_n). By comparing equations 3.19 and 3.20 it can be seen that the geometric and asperity failure components combine to give the roughness angle i. The basic and residual angles ϕ'_b, ϕ'_r are identical for unweathered rock, but where weathering or alteration has occurred in the joint walls, ϕ'_r may be less than ϕ'_b.

A simple empirical method of estimating ϕ'_r from ϕ'_b has been developed using the Schmidt hammer:

$$\phi'_r = \left(\phi'_b - 20\right) + 20\frac{r}{R} \qquad (3.21)$$

where R is the rebound on unweathered dry rock and r is the rebound on the weathered, saturated joint wall.

The need for equation 3.21 arises because simple tests can be used to find ϕ'_b, in which a smooth core stick with a similar core stick resting on it is tilted until sliding of the upper core stick occurs. A similar test, with a longitudinal joint in the core, gives a tilt α at sliding, from which JRC can be found using the expression

$$\text{JRC} = \frac{\alpha - \phi'_r}{\log_{10}\left(\text{JCS}/\sigma'_{n0}\right)} \qquad (3.22)$$

where σ'_{n0} is the effective stress normal to the joint due to the weight of the upper core Section. A Schmidt hammer test performed on the joint wall gives a simple empirical method for determining JCS.

As roughness affects small specimens to a greater degree than extensive joints in the field, it is necessary to apply scale corrections to the JRC given by tilt tests when using equation 3.22. Magnitudes of JRC range from zero for smooth planar surfaces to about 15 or 20 for very rough surfaces. A value of about 5 is appropriate for smooth almost planar surfaces such as many foliation and bedding surfaces, while a value of 10 is more appropriate for smooth undulating joints such as some bedding surfaces and smoother relief joints.

Barton and Choubey (1977) deduced from a study of 136 joint samples that the total friction angle (arctan τ_{jf}/σ'_n) could be estimated by the above methods to within about 1°, in the absence of scale effects. However, they concluded that scale effects were important in determining values of JCS and JRC, both values decreasing with increasing length of joint. Citing tests by Pratt *et al.* (1974) on quartz diorite which showed a drop in peak shear strength of nearly 40% for rough

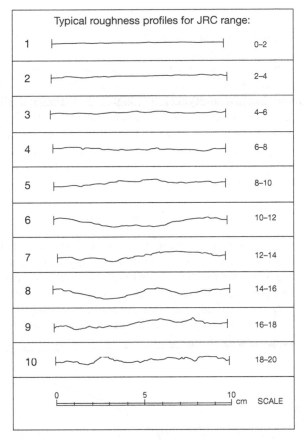

Figure 3.15 Roughness profiles and corresponding ranges of JRC values (ISRM, 1978). Reprinted from *International Journal of Rock Mechanics and Mining Sciences,* **15**, Bamford W.E. *et al.,* 'Suggested methods for the quantitative description of discontinuities in rock masses', pp. 319–368, 1978, with permission from Elsevier.

joint surfaces with areas ranging from 0.006 m² to 0.5 m², and adopting JRC = 20, they concluded that these tests indicated a fourfold reduction in JCR for a fivefold increase in joint length from 140 mm to 710 mm. Tilt tests by Barton and Choubey (1977) on rough planar joints of Drammen granite showed JRC values to decrease from 8.7 for 100 mm long joints to 5.5 for 450 mm long joints. ISRM (1978) recommendations for assessing joint roughness are based largely on the methods proposed by Barton and Choubey. Ten 'typical' roughness profiles, ranging from JRC = 0–2 to 18–20 and shown in Figure 3.15, are presented in this ISRM report. Based on limited laboratory tests, Barton and Bandis (1982) proposed a scale corrected JRC_n given by:

$$JRC_n = JRC_0(L_n/L_0)^{-0.02JRC_0} \tag{3.23}$$

where JRC_0 is the laboratory value and L_0, L_n refer to laboratory scale (100 mm) and *in situ* joint lengths respectively.

Drawing an analogy between coastal irregularities and rock joint roughnesses, fractal geometry (Mandelbrot, 1983) has been suggested as a means of reducing the degree of empiricism in quantifying joint roughness. A straight line linking the end points of a joint with length L can be divided into a number N_s of equal chords each of length r $(= L/N_s)$. If this chord length is stepped out along the joint profile (e.g. by means of dividers), the number N of segments it measures will be greater than N_s by an amount exceeding unity according to the roughness of the joint. The fractal dimension D, which is given by the expression:

$$D = -\log N/\log r \tag{3.24}$$

can be evaluated from the slope of a log–log plot of N against r. For practical usage it is necessary to convert this value into a quantity which can be used in design or analysis and one approach has been to try to establish a partly or wholly empirical relationship between JRC and fractal dimension (Turk *et al.*, 1987; Lee *et al.*, 1990). Lee *et al.* established an empirical expression relating the two, giving the plot shown in Figure 3.16. It will be noted that fractal dimension values for rock joints are not greatly in excess of unity and it is necessary to determine

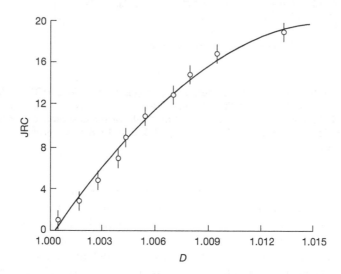

Figure 3.16 Plot of JRC range against fractal dimension D (Lee *et al.*, 1990). Reprinted from *International Journal of Rock Mechanics and Mining Sciences*, **27**, Lee, Y.H., Carr, J.R., Barr, D.J. and Haas, C.J., 'The fractal dimension as a measure of the roughness of rock discontinuity profiles', pp. 453–464, 1990, with permission from Elsevier.

the value to at least four decimal places. Nevertheless the value is operator dependent, as shown by published values of 1.0045 (Turk *et al.*, 1987), 1.005641 (Lee *et al.*, 1990) and 1.0040 (Seidel and Haberfield, 1995) for the 'standard' ISRM joint JRC = 10–12.

Adopting a more fundamental approach, and assuming gaussian distribution of chord angles θ, Seidel and Haberfield (1995) derived a relationship between fractal dimension and the standard deviation of asperity angle s_θ:

$$s_\theta \approx \cos^{-1}(N^{(1-D)/|D|}) \tag{3.25}$$

which plots as shown in Figure 3.17. It can be seen in Figure 3.17 that s_θ increases with D, which expresses the divergence from a straight line, and with N, the number of opportunities for divergence. As asperity height is a function of chord length and angle, an expression for the standard deviation of asperity height can be derived from equation 3.25:

$$s_h \approx (N^{-2/D} - N^{-2})^{0.5} \tag{3.26}$$

By generating random profiles for s_θ values and comparing these to the ISRM profiles for different ranges of JRC in Figure 3.15, Seidel and Haberfield (1995) established a subjective basis for predicting JRC from fractal geometry and showed, in addition, the possibility of the fractal approach providing conceptual models for the effects of normal stress on shear behaviour of joints and the scale dependence of joints.

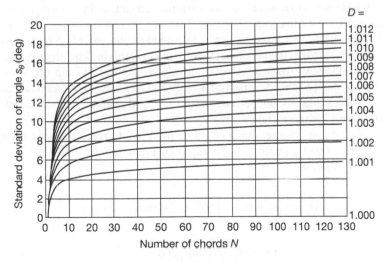

Figure 3.17 Variation of standard deviation of asperity angle s_θ with number of chords N and fractal dimension D given by equation 3.25 (Seidel and Haberfield, 1995). Reproduced by kind permission of Springer-Verlag.

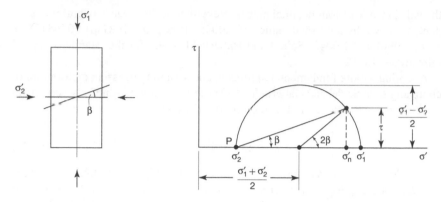

Figure 3.18 Magnitude of τ and σ'_n on a smooth planar joint.

The magnitude of σ'_n and τ acting on a smooth planar joint can be found from the principal biaxial stresses σ'_1, σ'_2 acting in the mass, as shown by the Mohr stress circle in Figure 3.18, which yields the equations

$$\sigma'_n = \tfrac{1}{2}\left(\sigma'_1 + \sigma'_2\right) + \tfrac{1}{2}\left(\sigma'_1 - \sigma'_2\right)\cos 2\beta \qquad (3.27a)$$

$$\tau = \tfrac{1}{2}\left(\sigma'_1 - \sigma'_2\right)\sin 2\beta \qquad (3.27b)$$

where β is the angle between the plane on which σ'_1 acts and the joint plane.

Barton has shown that this simple transformation can lead to substantial errors if dilation occurs during joint slippage. This can be seen in Figure 3.19, taken from Barton (1986). Biaxial tests were performed as shown in Figure 3.19, from which it was found that the resulting stress path (1), applying equations 3.27a and 3.27b, climbed above the peak strength envelope determined from full scale tilt tests. Part of this enhanced resistance was found to arise from frictional forces acting on the test platens, but correction for these still resulted in a stress path (2) climbing above the peak strength envelope.

It was found that the stress path could be modified to reach the peak strength envelope (stress path 3) without violating it by including the mobilized dilation angle i_{mob} in equations 3.27a and 3.27b to give

$$\sigma'_n = \tfrac{1}{2}\left(\sigma'_1 + \sigma'_2\right) + \tfrac{1}{2}\left(\sigma'_1 - \sigma'_2\right)\cos\left[2\left(\beta + i_{mob}\right)\right] \qquad (3.28a)$$

$$\tau = \tfrac{1}{2}\left(\sigma'_1 - \sigma'_2\right)\sin\left[2\left(\beta + i_{mob}\right)\right] \qquad (3.28b)$$

where

$$i_{\text{mob}} = \tfrac{1}{2}(\text{JRC})_{\text{mob}} \log_{10}\left(\frac{\text{JCS}}{\sigma'_n}\right) \qquad (3.29)$$

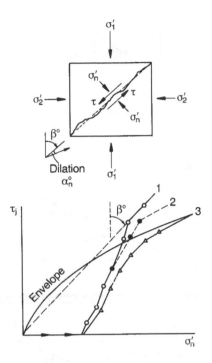

Figure 3.19 Biaxial stress paths on jointed rock specimen: 1, theoretical; 2, dilation
corrected; 3, fully corrected. (After Barton, 1986.)

If the dilation angle is not included in the stress transformation the applied stresses
required to cause failure may be significantly higher than estimated values. Barton
also suggests that stability analyses performed in plane strain environments
may give insufficient credit to the potential strength and stress changes caused
by slip of non-planar joints.

3.7 Influence of discontinuities in laboratory test specimens

The strength of a cylindrical rock specimen tested under triaxial conditions in
the laboratory may be strongly influenced by discontinuities within it. Jaeger
(1960) considered the case of a rock with a specific fracture strength, containing
a plane of weakness at an angle β to the plane containing the major principal stress,
as shown in Figure 3.20(a). He assumed the discontinuity to have Coulomb
cohesive and friction components of strength.

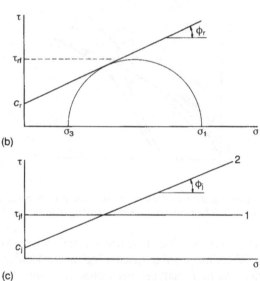

Figure 3.20 (a) Discontinuity in triaxial rock specimen; (b) intact rock strength; (c) two examples of joint strength (discussed in text).

It is a simple matter to use Mohr stress circles to show the influence of a planar discontinuity on the strength of a laboratory triaxial specimen.

Assume a rock specimen compressed axially, keeping the radial stress σ_3 constant, has intact strength parameters c_r, ϕ_r for the particular drainage conditions of the test as shown in Figure 3.20(b). At failure the shear strength for the specified applied value of σ_3 is τ_{rf}. For a rock with a curved envelope the strength parameters c_r, ϕ_r are the values given by the instantaneous tangent to the envelope. For fully drained conditions the relevant parameters will be c'_r, ϕ'_r, τ_{rf}.

Two possible cases for the strength of a discontinuity can be considered as shown in Figure 3.20(c).

1. Discontinuity has shear strength τ_{jf} independent of applied stress. This might be the case if the discontinuity is a joint with an infill of clay, in which excess pore pressures set up by the applied stresses cannot dissipate, either because the test is performed rapidly or because the joint walls confining the clay are impermeable.
2. Discontinuity has shear strength parameters c_j, ϕ_j, the magnitudes of which will be influenced by the rate of testing and nature of infill material (if any). The effective stress parameters c_j', ϕ_j', will apply if pore pressure in a clay infill can be dissipated fully during the test. For a clean joint or granular infill, c_j' will usually be zero and ϕ_j' only will be relevant.

Case I. Discontinuity with shear strength independent of applied stress

Referring to Figure 3.21(a), showing stress circles for a triaxial specimen tested under a lateral stress of σ_3, the application of axial stress σ_{1a} will not cause failure because the stress circles lie below the envelope for the discontinuity and for the intact rock. Applied axial stress σ_{1b} would cause failure of the specimen, even with no discontinuity present. Applied stress σ_{1c} produces a stress circle which lies partly above the τ_{jf} envelope, but wholly below the τ_{rf} envelope, which means that this stress cannot cause rupture of the intact rock, but may initiate sliding on a discontinuity, depending on the inclination of the discontinuity. In Figure 3.21(b), where stress circle c is reproduced, failure would not occur on discontinuities with inclination β_X or β_Z, because the stress conditions X and Z lie below the τ_{jf} envelope. On the other hand, failure would occur on a discontinuity with inclination β_Y, because stress condition Y lies above the τ_{jf} envelope.

In Figure 3.21(b), the strength envelope for discontinuities intersects the stress circle at points M and N. If there is any discontinuity in the test specimen with an inclination β to the horizontal, such that it intersects the stress circle between points M and N, then the specimen will fail under applied axial stress σ_{1c}. That is, failure will occur if $\beta_N \leqslant \beta \leqslant \beta_M$, where β_M and β_N are the inclinations of discontinuity directions PM and PN.

The limiting discontinuity inclinations $\beta_{max} = \beta_M$ and $\beta_{min} = \beta_N$, which would cause the specimen to fail under applied axial stress σ_{1c}, can be found by referring to Figures 3.21(c)–(e), and noting that

$$\tau_f = \tfrac{1}{2}\left(\sigma_{1c} - \sigma_{3c}\right) \tag{3.30}$$

From Figure 3.21(e):

$$\tau_{jf} = \tau_f \sin(180° - 2\beta)$$

$$\therefore \quad \tau_f = \frac{\tau_{jf}}{\sin 2\beta} \tag{3.31}$$

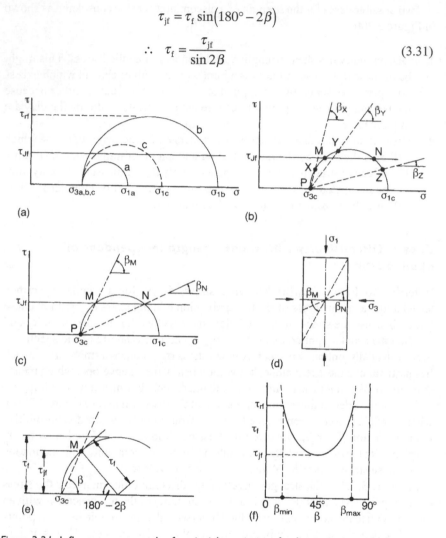

Figure 3.21 Influence on strength of a triaxial specimen of a discontinuity with shear strength independent of applied stress.

Equation 3.31 produces a plot of τ_f against β as shown in Figure 3.21(f), which is symmetrical about the angle $\beta = 45°$. As the strength of the specimen cannot exceed the intact strength of the rock τ_{rf}, this provides a cut-off for the plotted curve. It also means that discontinuities at an angle steeper than β_{max} or shallower than β_{min} have no influence on the strength of the specimen. These angles can be found by putting $\tau_f = \tau_{rf}$ in equation 3.31 to give

$$\sin 2\beta = \frac{\tau_{jf}}{\tau_{rf}} \tag{3.32}$$

If, for example, $\tau_{jf}/\tau_{rf} = 0.2$, then $\beta_{min} = 5.8°$ and $\beta_{max} = 84.2°$.

It can also be seen that the minimum value of τ_f occurs with the discontinuity having an angle $\beta = 45°$, for which $\tau_f = \tau_{rf}$.

Case 2. Discontinuity with shear strength parameters c_j, ϕ_j

Referring to Figure 3.22(a) it can be seen that under the lateral stress σ_{3a}, the applied stress σ_{1a} will not cause failure of the specimen because it lies below the strength envelope for the discontinuity. σ_{1b} will cause failure of the specimen, even with no discontinuity present. Applied stress σ_{1c} produces a stress circle which lies partly above the discontinuity envelope, but wholly below the strength envelope for the intact rock. The stress conditions represented by points X and Z on discontinuities at angles β_X and β_Z will not cause failure of the specimen. Stress condition Y on a discontinuity at an angle β_Y is not possible as it exceeds the strength of the discontinuity.

The maximum and minimum discontinuity angles for applied axial stress σ_{1c} are β_M and β_N shown in Figures 3.22(c) and (d). These angles can be found by reference to Figure 3.22(e), from which the following expression is obtained:

$$\tau_f = \frac{\sin\phi_j\left(c_j\cot\phi_j + \sigma_3\right)}{\sin\left(2\beta - \phi_j\right) - \sin\phi_j} \tag{3.33}$$

It would be usual to assume effective stress parameters $(\sigma'_3, \phi'_j, c'_j)$ in applying equation 3.33, i.e.

$$\tau_f = \frac{\sin\phi'_j\left(c'_j\cot\phi'_j + \sigma'_3\right)}{\sin\left(2\beta - \phi'_j\right) - \sin\phi'_j} \tag{3.34}$$

Use of equation 3.34 would mean either conducting the test at a sufficiently slow rate for full dissipation of excess pore pressures to occur, or measuring pore pressures in the discontinuity during the test, which is very difficult. If the discontinuity contains uncemented granular material such as silt or sand, or if it contains a soft clay, the effective stress cohesion parameter will be very small or zero. Substituting $c'_j = 0$ into equation 3.34 gives

$$\tau_f = \frac{\sigma'_3\sin\phi'_j}{\sin\left(2\beta - \phi'_j\right) - \sin\phi'_j} \tag{3.35}$$

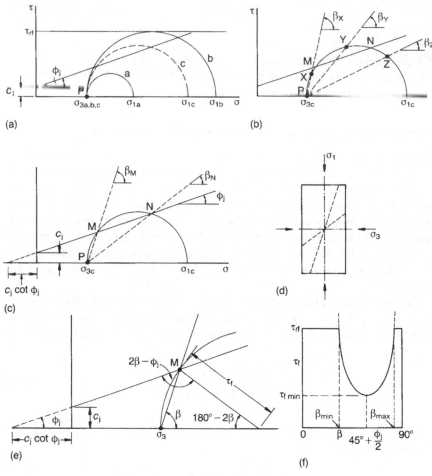

Figure 3.22 Influence on strength of a triaxial specimen of a discontinuity with shear
strength related to applied stress.

Equation 3.33 produces a plot of τ_f against β as shown in Figure 3.22(f).
By inspection of equation 3.33 it can be seen that the minimum value of τ_f occurs
when

$$\sin\left(2\beta - \phi_j\right) = 1$$

i.e.

$$\beta_{\text{crit}} = 45° + \frac{\phi_j}{2} \tag{3.36}$$

Substituting equation 3.36 into equation 3.33 gives

$$\tau_{f\,min} = \frac{\sin\phi_j\left(c_j\cot\phi_j + \sigma_3\right)}{1 - \sin\phi_j} \tag{3.37}$$

or, for effective stress parameters and $c'_j = 0$:

$$\tau_{f\,min} = \frac{\sigma'_3\sin\phi'_j}{1 - \sin\phi'_j} \tag{3.38}$$

The angles β_{max} and β_{min} can be found by substituting $\tau_f = \tau_{rf}$ into equation 3.33, thus

$$\tau_{rf} = \frac{\sin\phi_j\left(c_j\cot\phi_j + \sigma_3\right)}{\sin\left(2\beta - \phi_j\right) - \sin\phi_j} \tag{3.39}$$

Equation 3.39 yields the two angles $\beta_M = \beta_{max}$ and $\beta_N = \beta_{min}$. Discontinuities at steeper angles than β_{max} or shallower angles than β_{min} will not influence the strength of the test specimen.

For effective stress parameters, and putting $c'_j = 0$, equation 3.39 reduces to

$$\tau_{rf} = \frac{\sigma'_3\sin\phi'_j}{\sin\left(2\beta - \phi'_j\right) - \sin\phi'_j} \tag{3.40}$$

The general shape of the curves in Figure 3.22(f) have been confirmed in tests performed by Donath (1972) on a phyllite and by McLamore and Gray (1967) on a slate and two shales. Their test results are reproduced in Figure 3.23.

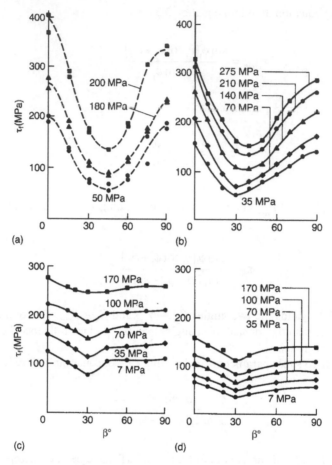

Figure 3.23 Variation of triaxial strength with orientation of plane of weakness for the confining pressures shown for: (a) Moretown phyllite; (b) slate; (c, d) two Green River shales. The broken lines in (a) are the present author's interpretation of the data. ((a) After Donath, 1972; (b to d) after McLamore and Gray, 1967.)

EXAMPLE 3.1 INFLUENCE OF DISCONTINUITY ON STRENGTH OF ROCK

The unconfined strength σ_c of an intact sandstone is 60 MPa, and under triaxial test conditions its strength can be represented by the expression

$$\sigma_{1N} = \left(\frac{M}{B} \sigma'_{3N} + 1 \right)^B$$

with values of $M = 12$ and $B = 0.6$. A test specimen placed in the triaxial cell under a cell pressure of 12 MPa contains a planar discontinuity with a slightly clayey sand infill, having effective stress strength parameters $c'_j = 0$, $\phi'_j = 30°$. An axial compression test is performed on the specimen at a rate which allows all dissipation of excess pore pressures. Assuming the discontinuity may have any inclination β to the horizontal, find the angle β_{crit} which gives the minimum strength and calculate the corresponding strength of the specimen. Find also the angles β_{max} and β_{min} above and below which the discontinuity will not affect the strength of the specimen.

Solution

$$\sigma'_{1N} = \left(\frac{M}{B} \sigma'_{3N} + 1 \right)^B$$

$$= \left(\frac{12}{0.6} \times \frac{12}{60} + 1 \right)^{0.6} = 2.63$$

$$\sigma'_1 = \sigma'_{1N} \times 60 = 158 \text{ MPa}$$

$$\tau_{rf} = \frac{\sigma'_1 - \sigma'_3}{2} = \frac{158 - 12}{2} = 73 \text{ MPa}$$

Equation 3.36: $\beta_{crit} = 45° + \dfrac{\phi'_j}{2}$

$$= 45° + 15° = 60°$$

Equation 3.38: $\tau_{f\min} = \dfrac{\sigma'_3 \sin \phi'_j}{1 - \sin \phi'_j}$

$$= \frac{12 \times 0.5}{1 - 0.5}$$

$$= 12 \text{ MPa}$$

Equation 3.40: $\tau_{rf} = \dfrac{\sigma'_3 \sin \phi'_j}{\sin(2\beta - \phi'_j) - \sin \phi'_j}$

$$\therefore \quad 73 = \frac{12 \times 0.5}{\sin(2\beta - 30°) - 0.5}$$

$$\therefore \quad 2\beta - 30° = 35.6° \text{ or } 144.4°$$

$$\therefore \quad \beta_{max} = 87.2° \quad \beta_{min} = 32.8°$$

EXAMPLE 3.2 INFLUENCE OF SECOND DISCONTINUITY

The specimen in Example 3.1 may have a second planar discontinuity consisting of a relatively impermeable seam of claystone with a strength, independent of applied stress, of 30 MPa. If the two discontinuities can have any angle β_S (clayey sand) or β_C (claystone), plot shear strength τ_f for the test specimen against β for the same test conditions as given in Example 3.1. Calculate the maximum and minimum β_S and β_C values which affect the strength of the specimen, and the minimum values of the specimen strength.

Solution

Clayey sand seam

$$\text{Equation 3.35: } \tau_f = \sigma_3' \frac{\sin \phi_j'}{\sin\left(2\beta_S - \phi_j'\right) - \sin \phi_j'}$$

$$= 12 \times \frac{\sin 30°}{\sin\left(2\beta_S - 30°\right) - \sin 30°} \text{MPa}$$

This equation plots as shown in Figure 3.24(a) giving, as for Example 3.1, a minimum specimen shear strength of 12 MPa, when the angle of the discontinuity is 60° to the horizontal. Also $\beta_{S\max} = 87.2°$ and $\beta_{S\min} = 32.8°$.

Claystone seam

$$\text{Equation 3.31: } \tau_{jf} = \tau_f \sin 2\beta_C$$

$$\therefore \ 30 = \tau_f \sin 2\beta_C$$

This equation plots as shown in Figure 3.24(b), giving a minimum shear strength for the specimen of 30 MPa with the discontinuity at 45° to the horizontal. The angles $\beta_{C\max}$ and $\beta_{C\min}$ are given by equation 3.32.

$$\text{Equation 3.32: } \sin 2\beta = \frac{\tau_{jf}}{\tau_{rf}} = \frac{30}{73} = 0.41$$

$$\therefore \ \beta_{C\max} = 77.9° \quad \beta_{C\min} = 12.1°$$

Superimposition of the diagrams in Figures 3.24(a) and (b) gives the envelope shown in Figure 3.24(c) for minimum strengths of the specimen, assuming both types of discontinuity to be present with possible inclinations to the horizontal between 0° and 90°.

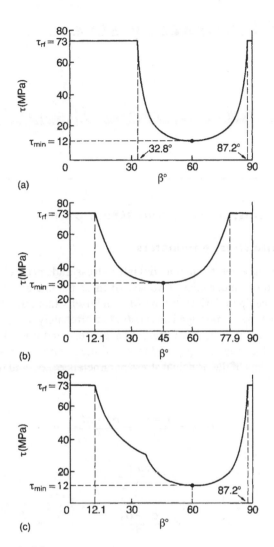

Figure 3.24 Example 3.2.

Chapter 4

Applied laboratory stress paths

4.1 Mohr circles, stresses and stress paths

4.1.1 Composite stress parameters

Composite stress parameters such as deviator stress and mean effective stress are widely used in soil mechanics. Two sets of these parameters are in common use, one having been popularized by the Massachusetts Institute of Technology (MIT) and the other by Cambridge University. Unfortunately, the two groups use the same symbols, q and p, but for different parameters, and it is intended here to adopt the Cambridge usage for q, p and to substitute t, s for q, p respectively, as used by MIT. In terms of the principal stress parameters σ_1, σ_3, and their effective stress equivalents:

$$t = t' = \frac{\sigma_1 - \sigma_3}{2} = \frac{\sigma_1' - \sigma_3'}{2} \tag{4.1}$$

$$s = \frac{\sigma_1 + \sigma_3}{2} \tag{4.2}$$

$$s' = \frac{\sigma_1' + \sigma_3'}{2} \tag{4.3}$$

The Cambridge parameters q, p are the stress invariants

$$q = \frac{1}{2^{1/2}} \left[\left(\sigma_1 - \sigma_2 \right)^2 + \left(\sigma_2 - \sigma_3 \right)^2 + \left(\sigma_3 - \sigma_1 \right)^2 \right]^{1/2} \tag{4.4}$$

$$p = \tfrac{1}{3} \left(\sigma_1 + \sigma_2 + \sigma_3 \right) \tag{4.5}$$

Under triaxial conditions, in compression, $\sigma_2 = \sigma_3$ and

$$q = q' = \sigma_1 - \sigma_3 = \sigma_1' - \sigma_3' \qquad (4.6)$$

$$p = \tfrac{1}{3}(\sigma_1 + 2\sigma_3) \qquad (4.7)$$

$$p' = \tfrac{1}{3}\left(\sigma_1' + 2\sigma_3'\right) \qquad (4.8)$$

Some difficulties arise in distinguishing between triaxial compression and triaxial extension where the directions of major and minor principal stresses interchange, and these can be overcome by the use of parameters σ_a, σ_r, for the axial and radial stresses respectively, thus:

$$q = q' = \sigma_a - \sigma_r = \sigma_a' - \sigma_r' \qquad (4.9)$$

$$p = \tfrac{1}{3}(\sigma_a + 2\sigma_r) \qquad (4.10)$$

$$p' = \tfrac{1}{3}(\sigma_a' + 2\sigma_r') \qquad (4.11)$$

The use of these parameters gives q positive for compression tests and negative for extension tests, thus allowing distinction between the two types of tests in plotting stress paths.

An important difference in the sets of parameters is the inclusion of the intermediate principal stress in the Cambridge set. While this is important in research and in establishing constitutive models for soils, the intermediate principal stress is usually not known. For this reason, and because of the direct relationship between t, s and the Mohr stress circles, it is t and s which are most used in practice.

The relationships between the stress parameters t, s and t', s' and the corresponding Mohr stress circles are shown by points X, X' in Figure 4.1.

X:
$$\frac{\sigma_a - \sigma_r}{2} = t \qquad \frac{\sigma_a + \sigma_r}{2} = s \qquad (4.12)$$

X':
$$\frac{\sigma_a' - \sigma_r'}{2} = t' \qquad \frac{\sigma_a' + \sigma_r'}{2} = s' \qquad (4.13)$$

These are the MIT stress parameters. It is possible to represent the stress circles simply by the points X, X' and thus each point on a t–s or t–s' stress path represents a stress circle.

The total stress and effective stress circles are separated by the pore pressure along the σ axis, but have the same diameter. Thus

$$s' = s - u \qquad (4.14)$$

$$t' = t \qquad (4.15)$$

In physical terms, equation 4.15 simply reflects the fact that the water in the voids is unable to offer any resistance to shear stress.

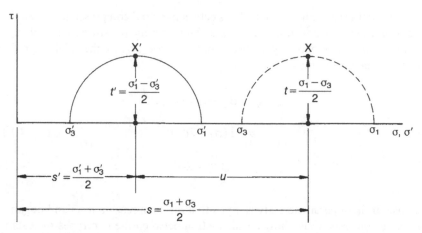

Figure 4.1 Relationships between stress parameters and Mohr circles. (Solid line, effective stress circle; broken line, total stress circle.)

4.1.2 Stress path plots

A stress path is a plot of a theoretical or experimental relationship between two stress parameters. Both total stress and effective stress paths can be plotted. It is possible simply to plot one principal stress against another, e.g. σ_a vs σ_r for triaxial tests, but it is more common to plot the relationship between composite stress parameters, i.e. t vs s, s' or q vs p, p'.

It has been shown in the previous Section that the t, s, s' parameters relate directly to the Mohr stress circle. Figure 4.2 shows a total stress circle and its corresponding effective stress circle. Point X has coordinates $\frac{1}{2}(\sigma_1 - \sigma_3)$ and $\frac{1}{2}(\sigma_1 + \sigma_3)$, i.e. t and s respectively. Similarly X' has coordinates $t'(= t)$, s'. The separation of X and X' is the pore pressure u_X.

If an increment of stress is applied, the stress conditions will change; as a simple example consider a conventional undrained triaxial compression test where σ_3 is held constant and σ_1 increased. The new circles will be such as those shown in Figure 4.2 with points Y, Y'. The total stress path is XY, the effective stress path is X'Y' and the pore pressure changes from u_X to u_Y.

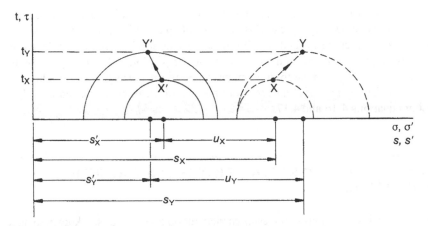

Figure 4.2 Relationships between stress paths and Mohr circles. (Solid line, effective stress circle; broken line, total stress circle.)

4.1.3 Failure stress conditions

Failure is usually taken as the point on a plot of deviator stress q against strain ε_a giving the maximum deviator stress. Other definitions are possible, but less common, such as the point of maximum stress ratio σ'_1/σ'_3, which may not necessarily coincide with maximum q.

At failure the Coulomb effective stress envelope is tangential to the effective stress circle, with slope ϕ' and intercept c', as shown in Figure 4.3. A failure line can also be drawn through X′ in Figure 4.1 which represents the failure envelope for t–s' stress paths. This is the K_f line and has a slope α' and intercept k'. It is easy to show that

$$\tan \alpha' = \sin \phi' \qquad (4.16)$$

$$k' = c'\cos \phi' \qquad (4.17)$$

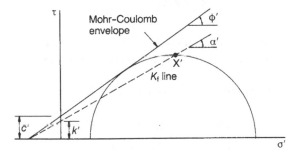

Figure 4.3 Mohr–Coulomb envelope and K_f line.

It can be seen in Figure 4.1 that the effective stress conditions at failure for a specific test can be fully represented by the single point X'. If a substantial number of tests have been performed, exhibiting some scatter of results, it may be easier to draw an envelope through such points on a t–s' plot, either by eye or using a regression analysis, than to draw an envelope to the corresponding circles on a τ–σ' plot. The values of k' and α' from the t–s' plot can be converted into c', ϕ' using equations 4.16 and 4.17.

EXAMPLE 4.1 DRAINED STRESS PATHS AND STRENGTH PARAMETERS

A drained triaxial compression test is carried out on a sample of soil known to have the effective stress strength characteristics $c' = 10$ kPa, $\phi' = 22°$. If the cell pressure is 100 kPa draw the Mohr stress circle at failure and evaluate the failure values of t, s', q and p'. Draw the stress paths on both t–s' and q–p' diagrams. What are the slopes of the stress paths? Evaluate k' and α'.

Solution

Referring to Figure 4.4(a), the radius of stress circle at failure is

$$r = (10\cot 22° + 100 + r)\sin 22° \text{ kPa}$$
$$\therefore r = 74.7 \text{ kPa}$$
$$t_f = r = 74.7 \text{ kPa}$$
$$s'_f = 100 + r = 174.7 \text{ kPa}$$
$$q_f = 2r = 149.4 \text{ kPa}$$
$$\sigma'_{1f} = 100 + 2r = 249.4 \text{ kPa}$$
$$p'_f = \tfrac{1}{3}(249.4 + 2 \times 100) \text{ kPa}$$
$$\therefore p'_f = 149.8 \text{ kPa}$$

The stress paths are shown in Figures 4.4(b) and (c). It can be seen that the t–s' stress path has a slope of 1:1 and the q–p' stress path has a slope of 3:1.

$$\text{Equation 4.17: } k' = 10 \cos 22° = 9.3 \text{ kPa}$$

$$\text{Equation 4.16: } \tan \alpha' = \sin \phi'$$
$$\therefore \alpha' = 20.5°$$

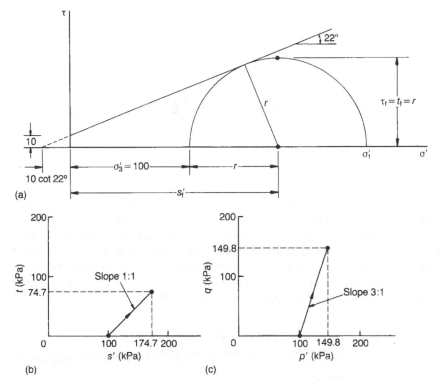

Figure 4.4 Example 4.1.

4.2 Consolidation stresses and stress paths

Consolidation in the laboratory may be carried out either to measure the consolidation characteristics of the soil or to impart a predetermined stress history and stress state to a soil specimen before submitting it to a shear test. The consolidation cell or oedometer is used for the former purpose, while the triaxial cell can be used for either purpose. Furthermore, it is possible in the triaxial cell to subject the test specimen to either isotropic stress or some other stress condition with a predetermined ratio of axial to radial effective stress. In the latter case the most common ratio selected is that which gives zero radial strain, and thus corresponds to the one-dimensional consolidation in the oedometer.

4.2.1 Isotropic consolidation

A sample of soil under isotropic stress has a Mohr stress circle of zero radius. If it has zero pore pressure it will be represented on a *t*, *s'* plot by a point such as A, A'

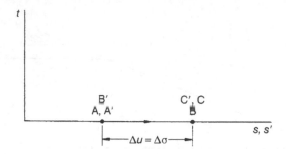

Figure 4.5 Stress paths for isotropic consolidation.

in Figure 4.5. A is the total stress point, A′ the effective stress point, and these are coincident. If an increment of isotropic stress $\Delta\sigma_a = \Delta\sigma_r = \Delta\sigma$ is now applied, it is instantaneously carried entirely by the pore water $\Delta u = \Delta\sigma$, and the effective stress path does not change. Thus, the sample follows total stress path A → B along the line $t = 0$ to point B. As excess pore pressure dissipates, the sample follows effective stress path B′ → C′, but the total stress remains at B = C. If the total stress increment $\Delta\sigma$ is now removed the stress paths will be reversed.

4.2.2 One-dimensional consolidation

The ratio between the radial and axial effective stresses during one-dimensional consolidation is given by

$$\sigma_r' = K_0\sigma_a' \tag{4.18}$$

For the particular case of a normally consolidated soil:

$$K_0 = K_{nc} < 1$$

and the state of stress in the test specimen is given by the stress circle in Figure 4.6(a). The slope of the K_{nc} line is

$$\frac{\Delta t}{\Delta s'} = \frac{1 - K_{nc}}{1 + K_{nc}} \tag{4.19}$$

The stress path representation of the stress circle in Figure 4.6(a) is given by the point D through which the K_{nc} line passes in Figure 4.6(b).

On application of a total stress increment $\Delta\sigma_a$ the instantaneous response is an equal change in pore pressure, and in σ_r. Thus

$$\Delta u = \Delta\sigma_r = \Delta\sigma_a = \Delta\sigma$$

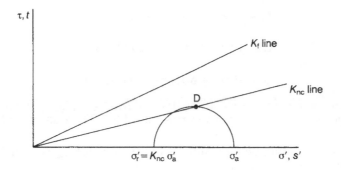

(a)

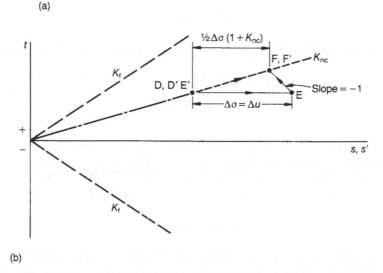

(b)

Figure 4.6 One-dimensional consolidation: (a) K_{nc} line; (b) total and effective stress paths.

The total stress path $D \rightarrow E$ is followed, but the effective stress condition E' remains at D'. In Figure 4.6(b),

$$DE = \Delta\sigma = \Delta u$$

On dissipation of pore pressure Δu the following stress changes occur:

$$\Delta\sigma_a = 0 \quad \Delta\sigma'_a = \Delta u \tag{4.20}$$

$$\text{Equation 4.18:} \ \Delta\sigma'_r = K_{nc}\Delta\sigma'_a = K_{nc} \, \Delta u \tag{4.21}$$

$$\text{Equation 1.32:} \ \Delta\sigma_r = -\Delta u + K_{nc}\Delta u = -\Delta u(1 - K_{nc}) \tag{4.22}$$

$$\Delta t = \tfrac{1}{2}(\Delta\sigma_a - \Delta\sigma_r) = \tfrac{1}{2}\Delta u(1 - K_{nc}) \qquad (4.23)$$

$$\Delta s = \tfrac{1}{2}(\Delta\sigma_a + \Delta\sigma_r) = -\tfrac{1}{2}\Delta u(1 - K_{nc}) \qquad (4.24)$$

$$\Delta s' = \tfrac{1}{2}(\Delta\sigma'_a + \Delta\sigma'_r) = \tfrac{1}{2}\Delta u(1 + K_{nc}) \qquad (4.25)$$

The corresponding total and effective stress paths EF, E'F' are shown in Figure 4.6(b).

For most practical purposes it is sufficient to assume that K_{nc} is given by the modified Jâky (1944) expression:

$$K_{nc} = 1 - \sin\phi' \qquad (4.26)$$

and substituting equation 4.26 into equation 4.19 gives the slope of the K_{nc} line as

$$\frac{\Delta t}{\Delta s'} = \frac{\sin\phi'}{2 - \sin\phi'} \qquad (4.27)$$

If a normally consolidated soil at a point such as F, F' in Figure 4.6(b) is subject to a decrease in axial stress $-\Delta\sigma_a$, a reversal of effective stress path along the K_{nc} line is not followed. Instead, a path such as F'G'H' is followed, as shown in Figure 4.7, and the ratio of stress σ'_r/σ'_a at H' is greater than K_{nc}. The ratio

$$K_0 = \sigma'_r / \sigma'_a$$

for a soil depends on the overconsolidation ratio (OCR) and soil type. Brooker and Ireland (1965) conducted instrumented oedometer tests on a range of soils and found the relationship between K_0 and OCR to vary with plasticity index in the manner shown in Figure 4.8. The amount of experimental evidence of variation of K_0 with OCR is limited, but Nadarajah (1973) and Clegg (1981) present test results for kaolin ($w_L = 72\%$, $w_p = 40\%$) from instrumented oedometer tests. The measured variation of K_0 with OCR given by Clegg is shown in Figure 4.9.

A number of expressions relating K_0 to OCR have been presented. Assuming isotropic elastic behaviour during swelling, and zero lateral strain $\Delta\varepsilon'_r$, Wroth (1975) showed that for a soil with effective stress elastic modulus E' and Poisson's ratio v':

$$0 = E'\,\Delta\varepsilon_r = \Delta\sigma'_r - v'\,\Delta\sigma'_r - v'\,\Delta\sigma'_a$$

$$\therefore \quad \Delta\sigma'_r = \frac{v'}{1 - v'}\Delta\sigma'_a \qquad (4.28)$$

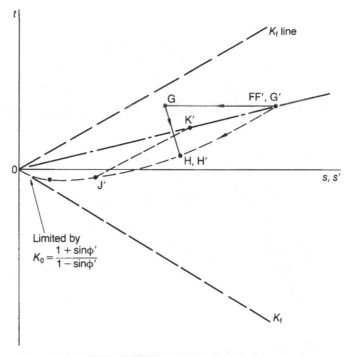

Figure 4.7 Total and effective stress paths followed during one-dimensional overconsolidation.

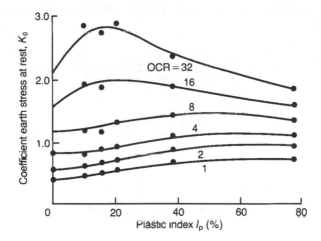

Figure 4.8 Relationships between K_0, OCR and plasticity index. (After Brooker and Ireland, 1965.)

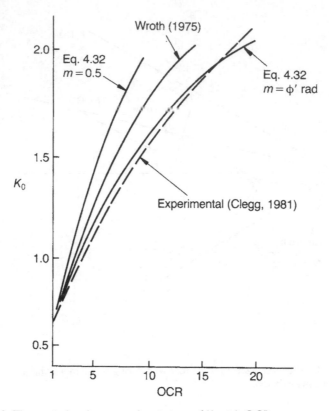

Figure 4.9 Theoretical and measured variations of K_0 with OCR.

If the stresses on the normally consolidated soil before swelling are σ'_a, σ'_r and $K_{nc} = \sigma'_r/\sigma'_a$:

$$K_0 = \frac{\sigma'_r - \Delta\sigma'_r}{\sigma'_a - \Delta\sigma'_a} \tag{4.29}$$

$$OCR = \frac{\sigma'_a}{\sigma'_a - \Delta\sigma_a} \tag{4.30}$$

Thus

$$K_0 = OCR \cdot K_{nc} - \frac{v'}{1-v'}(OCR - 1) \tag{4.31}$$

where Poisson's ratio v' varies with plasticity index I_p, as shown in Figure 4.10 after Wroth (1975). This equation matches experimental data reasonably well for

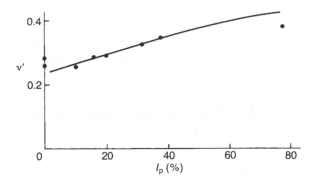

Figure 4.10 Relationship between Poisson's ratio and plasticity index. (After Wroth, 1975.)

lightly overconsolidated soils, but not for heavily overconsolidated soils, and Wroth (1975) presents a more complex equation to represent this condition.

A number of workers have suggested that K_0 is related to OCR by an expression of the form

$$K_0 = K_{nc}(OCR)^m \qquad (4.32)$$

with m values ranging from 0.41 (Schmertmann, 1975) to 0.5 (Meyerhof, 1976). Schmidt (1966) suggested

$$m = 1.25 \sin \phi'$$

The present author prefers the use of

$$m = \phi' \text{ radians}$$

Figure 4.9 shows the experimental data for kaolin, together with various theoretical predictions.

The experimental data in Figure 4.9 give stress paths of the form G′ H′ J′ 0 shown in Figure 4.7. A limitation is placed on the maximum value of K_0 because the stress path cannot cross the extension failure envelope. Thus, the maximum value of K_0 if $c' = 0$ is given by the expression

$$K_{0max} = \frac{1 + \sin \phi'}{1 - \sin \phi'} \qquad (4.33)$$

Inclusion of c' would increase this value; but even with $c' = 0$, higher values than expected might occur because ϕ' is the extension value, which may be several degrees higher than the compression value.

If heavily overconsolidated soil is reloaded, the stress path returns quickly to the K_{nc} condition shown by J′ K′ in Figure 4.7, so that $K_0 = K_{nc}$ is achieved while the sample is still overconsolidated. Burland and Hancock (1977) assuming isotropic elastic behaviour for a stiff clay in the field applied equation 4.28 to show that

$$\frac{\Delta \sigma'_h}{\Delta \sigma'_v} = \frac{v'}{1-v'} = 0.18 \text{ if } v' = 0.15$$

Allowance for anistropy might increase this value, but it is clear that the ratio is small and would predict a rapid return to the K_{nc} condition.

This behaviour has been shown experimentally for kaolin in an instrumented oedometer by Manson (1980), who performed reload loops at different over-consolidation ratios. The stress path plots are shown in Figure 4.11.

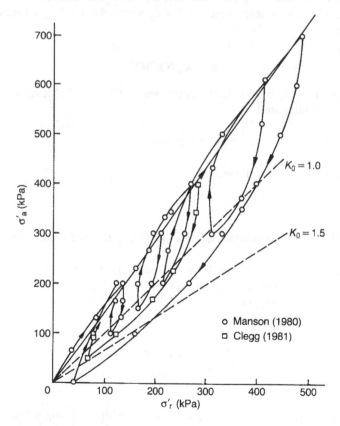

Figure 4.11 Cyclic unload–reload one-dimensional consolidation loops for Speswhite kaolin.

EXAMPLE 4.2 CONSOLIDATION STRESS PATHS

It is possible to consolidate a soil specimen in the triaxial cell by axial compression, keeping the diameter effectively constant using a device which is able to detect very small radial strains in the specimen. If a sample of clay, reconstituted at a high moisture content, is consolidated in this way under 80 kPa increments of axial stress up to 480 kPa, then allowed to swell, still maintaining zero radial strain conditions, under five axial stress decrements of 80 kPa and a final decrement of 40 kPa, what is the final OCR of the specimen? Plot the stress path on a t–s' diagram, assuming $\phi' = 23.6°$ and m in equation 4.32 is equal to ϕ' in radians. If the specimen is now reconsolidated under increasing axial stress, what would be the axial pressure when the stress ratio σ'_r/σ'_a again equalled K_{nc}, assuming isotropic elastic behaviour?

Solution (see Table 4.1)

$$\text{Equation 4.26: } K_{nc} = 1 - \sin 23.6° = 0.60$$

$$\text{Equation 4.32: } K_0 = 0.60(\text{OCR})^{0.41}$$

The stress path plot (from the data in Table 4.1) is shown in Figure 4.12.

Table 4.1 Calculated stress path data for Example 4.2

Axial stress	Normally consolidated				Overconsolidated			
σ'_a (kPa)	σ'_r (kPa)	t (kPa)	s' (kPa)	OCR	K_0	σ'_r (kPa)	t (kPa)	s' (kPa)
40	–	–	–	12	1.66	66	−13	53
80	48	16	64	6	1.25	100	−10	90
160	96	32	128	3.0	0.94	150	10	155
240	144	48	192	2.0	0.80	192	24	216
320	192	64	256	1.5	0.71	227	46	274
400	240	80	320	1.2	0.65	260	70	330
480	288	96	384	1.0	0.60	288	96	384

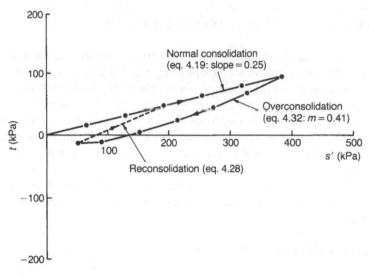

Figure 4.12 Example 4.2.

Reconsolidation:

Equation 4.28:
$$\frac{\Delta\sigma'_r}{\Delta\sigma'_a} = \frac{v'}{1-v'} = \frac{0.25}{0.75} = 0.33$$

$$\frac{66 + \Delta\sigma'_r}{40 + \Delta\sigma'_a} = K_{nc}$$

$$\therefore \frac{66 + 0.33\Delta\sigma'_a}{40 + \Delta\sigma'_a} = 0.60$$

$$\therefore \Delta\sigma'_a = 156 \text{ kPa}$$

$$\therefore \sigma'_a = 196 \text{ kPa}$$

The broken line in Figure 4.12 shows the reconsolidation stress path.

4.3 Drained triaxial stress paths

Although a triaxial test is restricted in so far as the intermediate principal stress must be the cell pressure, and therefore must equal the minor or major principal stress, nevertheless a wide range of drained stress paths is possible because the axial pressure σ_a and radial pressure σ_r can be varied independently. Either may be increased, decreased or held constant.

4.3.1 Drained compression test with constant radial stress (DI)

In a drained triaxial test the effective and total stresses are the same, i.e. $\sigma_a = \sigma'_a$ and $\sigma_r = \sigma'_r$, assuming pore pressure datum is zero. If stresses are initially isotropic and pore pressure zero, the stress path will have an initial point such as M in Figure 4.13.

The only stress which changes is σ'_a by an amount $\Delta\sigma'_a$ i.e.

$$\Delta\sigma'_a = +\Delta\sigma'_a$$
$$\Delta\sigma'_r = 0$$
$$\Delta t = \frac{\Delta\sigma'_a - \Delta\sigma'_r}{2} = +\frac{\Delta\sigma'_a}{2} \qquad (4.34)$$
$$\Delta s' = \frac{\Delta\sigma'_a + \Delta\sigma'_r}{2} = +\frac{\Delta\sigma'_a}{2}$$
$$\therefore \frac{\Delta t}{\Delta s'} = +1$$

and the stress path is a straight line with slope 1:1, as shown in Figure 4.13, failing at P. Samples with initial stress conditions such as N would also fail at P.

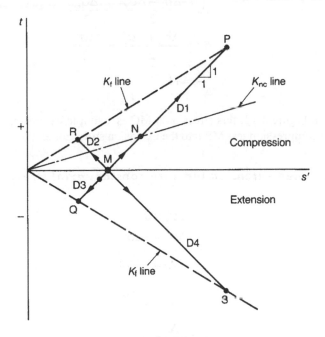

Figure 4.13 Drained triaxial stress paths for initial $K_0 = 1$.

4.3.2 Drained compression test with constant axial stress (D2)

$$\Delta \sigma'_a = 0$$
$$\Delta \sigma'_r = -\Delta \sigma'_r$$
$$\Delta t = +\frac{\Delta \sigma'_r}{2}$$
$$\Delta s' = -\frac{\Delta \sigma'_r}{2}$$
$$\therefore \quad \frac{\Delta t}{\Delta s'} = -1$$

(4.35)

If the stresses are initially isotropic the resulting stress path is given by MR in Figure 4.13.

4.3.3 Drained extension test with constant radial stress (D3)

$$\Delta \sigma'_a = -\Delta \sigma'_a$$
$$\Delta \sigma'_r = 0$$
$$\Delta t = \frac{\Delta \sigma'_a - \Delta \sigma'_r}{2} = -\frac{\Delta \sigma'_a}{2}$$
$$\Delta s' = \frac{\Delta \sigma'_a + \Delta \sigma'_r}{2} = -\frac{\Delta \sigma'_a}{2}$$
$$\therefore \quad \frac{\Delta t}{\Delta s'} = +1$$

(4.36)

As seen in Figure 4.13, this stress path MQ is a simple extension of the corresponding compression test MP into the extension stress zone.

4.3.4 Drained extension test with constant axial stress (D4)

$$\Delta \sigma'_a = 0$$
$$\Delta \sigma'_r = +\Delta \sigma'_r$$
$$\Delta t = -\frac{\Delta \sigma'_r}{2}$$
$$\Delta s' = +\frac{\Delta \sigma'_r}{2}$$
$$\therefore \quad \frac{\Delta t}{\Delta s'} = -1$$

(4.37)

As seen in Figure 4.13, this stress path MS is a simple projection of the corresponding compression test path MR into the extension stress zone.

4.4 Influence of stress paths on laboratory-measured drained strengths

It can be seen in Figure 4.13 that the failure strengths in drained tests are greatly influenced by the stress paths. Considering the simple case of test specimens with identical initial isotropic stress, the following relationships for the magnitudes of failure strengths follow from the geometry of Figure 4.13:

$$t_f(D1) = t_f(D4) \qquad (4.38)$$

$$t_f(D2) = t_f(D3) \qquad (4.39)$$

$$\frac{t_f(D1)}{t_f(D2)} = \frac{t_f(D4)}{t_f(D3)} = \frac{\sin(45° + \phi')}{\sin(45° - \phi')} \qquad (4.40)$$

For $\phi' = 20°$ and $30°$ the strength ratios given by equation 4.40 are, respectively, 2.1 and 3.7.

Considering triaxial compression stress paths with constant radial stress and constant axial stress respectively, the influence of c' and initial stress conditions can be shown by referring to Figures 4.14 and 4.15.

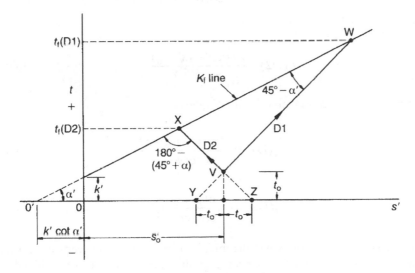

Figure 4.14 Drained triaxial stress paths for initial $K_0 < 1$

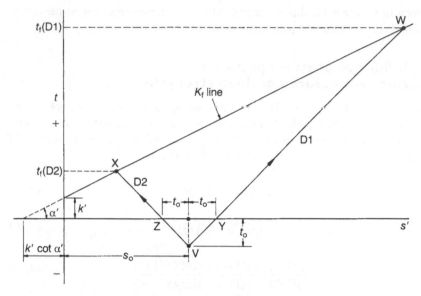

Figure 4.15 Drained triaxial stress paths for initial $K_0 > 1$.

Figure 4.14 depicts the case where the initial stresses represented by point V are t_0, s_0' and the vertical stress exceeds the horizontal stress. It can be deduced from the triangles $0'WY$ and $0'XZ$ respectively that

$$t_f(D1) = \frac{k'\cot\alpha' + (s_0' - t_0)}{\cot\alpha' - 1} \tag{4.41}$$

and

$$t_f(D2) = \frac{k'\cot\alpha' + (s_0' + t_0)}{\cot\alpha' + 1} \tag{4.42}$$

Putting $k' = c'\cos\phi'$ and $\tan\alpha' = \sin\phi'$ gives

$$t_f(D1) = \frac{c'\cot\phi' + (s_0' - t_0)}{\operatorname{cosec}\phi' - 1} \tag{4.43}$$

$$t_f(D2) = \frac{c'\cot\phi' + (s_0' + t_0)}{\operatorname{cosec}\phi' + 1} \tag{4.44}$$

These equations also hold when the initial horizontal stress exceeds the vertical stress as depicted in Figure 4.15, but the t_0 values in equations 4.41 to 4.44 will be negative.

In routine laboratory triaxial tests it is not usual to attempt to set up test specimens under initial field stresses in order to measure strength values. After extrusion of the test specimen from the sampling tube, it is under zero (and hence isotropic) external stress. In fine-grained soils the integrity of the sample is maintained after extrusion by negative pore pressure u_e which induces isotropic effective stress s'_e in the soil. Thus

$$s'_e = -u_e \qquad (4.45)$$

The magnitude of u_e is governed by the vertical and horizontal effective stresses in the ground σ'_{v0} and σ'_{h0}, usually taken to be principal stresses. In the absence of better information, an estimate of u_e can be made by assuming the soil to behave as an undrained isotropic elastic medium during removal of the field stresses (i.e. during the sampling and extruding process). Experience with a particular soil might of course indicate a better basis of prediction. Elastic behaviour of soil is discussed fully in Chapter 5, and it is sufficient to say here that if undrained isotropic elastic behaviour obtains, then for any applied stress change, the change in mean effective stress is zero. Thus

$$u_e = -\tfrac{1}{3}(\sigma'_{v0} + 2\sigma'_{h0}) \qquad (4.46)$$

and

$$s'_e = \tfrac{1}{3}(\sigma'_{v0} + 2\sigma'_{h0}) \qquad (4.47)$$

Drained tests in the laboratory are usually performed to determine the c', ϕ' envelope, rather than to obtain a specific value of t_f. Normally at least three tests are performed to locate points on the envelope, and the precise starting stresses may not be important as long as the portion of the envelope located corresponds to that which is relevant to the field problem, and the stress paths bear some relationship to the field problem.

In performing drained triaxial tests, then, each specimen will be set up under a cell pressure σ_{cp} and the initial pore pressure equal to $(\sigma_{cp} + u_e)$ will be allowed to dissipate before shearing the specimen, resulting in an initial isotropic effective stress s'_i equal in magnitude to σ_{cp}. The corresponding drained strengths will be, from equations 4.43 and 4.44,

$$t_f(\text{D1}) = \frac{c' \cot \phi' + s'_i}{\operatorname{cosec} \phi' - 1} \qquad (4.48)$$

$$t_f(\text{D2}) = \frac{c' \cot \phi' + s'_i}{\operatorname{cosec} \phi' + 1} \qquad (4.49)$$

EXAMPLE 4.3 INITIAL PORE PRESSURE AND DRAINED STRENGTHS FOR A SOFT CLAY

Undisturbed soil specimens are taken from a depth $z = 5$ m in a soft, lightly overconsolidated clay for which $K_0 = 0.7$, unit weight $\gamma = 16$ kN/m³, $c' = 0$, ϕ' = 22°. The water table is at a depth $z_w = 1$ m, and assume $\gamma_w = 10$ kN/m³. Estimate u_e and find $t_f(D1)$, $t_f(D2)$ for specimens tested under a cell pressure of 40 kPa. (Assume $\sigma_{v0} = \gamma z$.)

Solution

$$\sigma_{v0} = \gamma z = 80 \text{ kPa}$$
$$\sigma'_{v0} = \gamma z - \gamma_w(z - z_w) = 40 \text{ kPa}$$
$$\sigma'_{h0} = K_0 \sigma'_{v0} = 28 \text{ kPa}$$

Equation 4.46: $u_e = -32$ kPa

Equation 4.48: $t_f(D1) = 24$ kPa

Equation 4.49: $t_f(D2) = 11$ kPa

EXAMPLE 4.4 INITIAL PORE PRESSURE AND DRAINED STRENGTHS FOR A STIFF CLAY

Undisturbed samples are taken from a depth of 5 m in a stiff, heavily overconsolidated clay for which $K_0 = 1.8$, $\gamma = 20$ kN/m³, $c' = 5$ kPa, $\phi' = 22°$. The water table is at a depth of 3 m and assume $\gamma_w = 10$ kN/m³. Estimate u_e and find $t_f(D1)$, $t_f(D2)$ for specimens tested under a cell pressure of 100 kPa. (Assume $\sigma_{v0} = \gamma z$.)

Solution

$$\sigma_{v0} = 100 \text{ kPa}, \quad \sigma'_{v0} = 70 \text{ kPa}, \quad \sigma'_{h0} = 126 \text{ kPa}$$

Equation 4.46: $u_e = -107$ kPa

Equation 4.48: $t_f(D1) = 67$ kPa

Equation 4.49: $t_f(D2) = 30$ kPa

4.5 Undrained triaxial stress paths

4.5.1 Undrained compression test with constant radial total stress (U1)

The slopes of the total stress paths in conventional undrained compression are found by putting $\Delta\sigma_r = 0$ and thus $\Delta t = \frac{1}{2}\Delta\sigma_a$, $\Delta s = \frac{1}{2}\Delta\sigma_a$ and $\Delta t/\Delta s = 1$. The effective stress paths will be separated from these by the pore pressure value u at any time.

Skempton (1948) presented the equation

$$\Delta u = B[\Delta\sigma_3 + A(\Delta\sigma_1 - \Delta\sigma_3)] \qquad (4.50)$$

with $B = 1$ for a saturated clay:

$$\Delta u = A\Delta\sigma_1 + \Delta\sigma_3(1 - A) \qquad (4.51)$$

where A at failure typically has values 0.33 to 1 for lightly overconsolidated or normally consolidated clay and 0 to -0.25 for heavily overconsolidated clay.

As $\Delta\sigma_3 = \Delta\sigma_r = 0$, it can be seen from equation 4.51 that

$$\Delta u = A\Delta\sigma_1 = 2A\Delta t \qquad (4.52)$$

If A is constant during the test, the effective stress path is a straight line with slope, as seen in Figure 4.16, of

$$\frac{\Delta t}{\Delta s'} = \frac{1}{1 - 2A} \qquad (4.53)$$

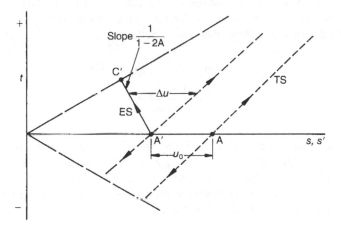

Figure 4.16 Undrained triaxial total (TS) and effective (ES) stress paths (assumed linear).

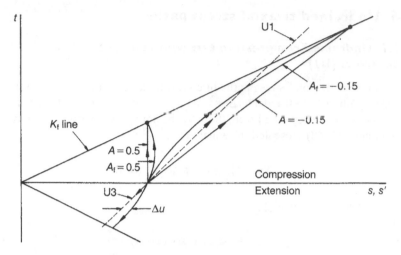

Figure 4.17 Typical undrained triaxial effective stress paths for $A_f = 0.5$ and $A_f = -0.15$. Broken lines indicate total paths.

Effective stress t–s' paths for $A = 0.5$ and -0.15 are shown in Figure 4.17. In practice A is usually not constant through a test, and more typical stress paths are also shown in Figure 4.17 for lightly overconsolidated clay (LOC), for which $A_f = 0.5$ at failure, and a heavily overconsolidated clay (HOC), for which $A_f = -0.15$ at failure.

4.5.2 Undrained compression test with constant axial total stress (U2)

A simple triaxial compression test can be carried out by holding the axial total stress constant and decreasing the lateral total pressure. This gives a total stress undrained path with a slope obtained by putting

$$\Delta\sigma_a = 0 \qquad \Delta\sigma_r = -\Delta\sigma_3$$

$$\frac{\Delta t}{\Delta s} = \frac{0 + \Delta\sigma_3}{0 - \Delta\sigma_3} = -1 \tag{4.54}$$

In an undrained test carried out in this manner the effective stress path is the same as for a conventional test (given by equation 4.53), the explanation for which is seen in Figure 4.18. The stress condition C can be obtained from A by simply subtracting B, which is an isotropic total stress change causing no change in effective stress. Thus, in effective stress terms,

$$A = B + C$$

but as B causes no change in effective stress,

$$A = C$$

The total stress paths are different, and consequently the pore pressures developed are different, as seen in Figure 4.19. If, on the application of Δt in a conventional test, a pore pressure Δu_1 is developed, then in the test, with decreasing lateral pressure the pore pressure

$$\Delta u_2 = 2\,\Delta t - \Delta u_1$$

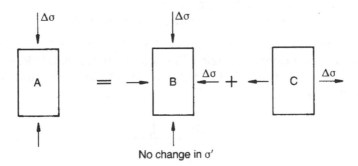

Figure 4.18 Illustration of uniqueness of effective stress paths for U1 and U2 tests on identical specimens.

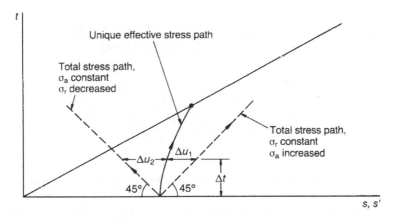

Figure 4.19 Relative pore pressures developed in U1 and U2 tests on identical specimens.

4.5.3 Undrained extension tests (U3 and U4)

Starting from an isotropic stress condition, σ_r is the major principal stress and σ_a is the minor principal stress throughout the test. The tests may be performed by holding the radial total stress constant and decreasing the axial total stress σ_a(U3), or by holding the axial total stress constant and increasing the radial total stress σ_r(U4). These tests give identical effective stress paths for identical soil specimens. Equation 4.50 breaks down for these tests, but as shown in Figure 4.15 the difference between the total stress and effective stress paths is still given by the pore pressure.

4.6 Influence of stress paths on laboratory-measured undrained strengths

It follows from the discussion in Section 4.5.1 that for undrained triaxial compression tests the slope of a straight line connecting the initial effective stress condition of a specimen with the point where the effective stress path meets the K_f line is given by

$$\frac{\Delta t}{\Delta s'} = \frac{1}{1-2A_f} \tag{4.55}$$

Referring to Figure 4.20 it can be seen that, for a soil with initial stress conditions t_0, s'_0 the failure strength t_f(UC) is unique for all undrained triaxial compression tests. Thus

$$t_f(U1) = t_f(U2) = t_f(UC) = \frac{k' + s'_0 \tan\alpha' - (1-2A_f)t_0 \tan\alpha'}{1-(1-2A_f)\tan\alpha'} \tag{4.56}$$

or

$$t_f(UC) = \frac{c' \cot\phi' + s'_0 - (1-2A_f)t_0}{\mathrm{cosec}\,\phi' - 1 + 2A_f} \tag{4.57}$$

However, as discussed in Section 4.6, an undisturbed fine grained specimen immediately after extrusion from a sampling tube will be under zero (isotropic) external stress and have a negative pore pressure $-u_e$. This gives rise to an isotropic effective stress $s'_e = -u_e$ where, in the absence of better information, an estimate of u_e can be made by assuming equation 4.46 to hold.

If the test specimen is saturated, the application of a cell pressure σ_{cp} will cause a corresponding increase in pore pressure $\Delta u = \sigma_{cp}$, but the effective stress will remain equal to s'_e. Thus

$$s_i' = s_e'$$

Consequently the undrained compression strength $t_f(UC)$ is given by putting $t_0 = 0$ and $s_0' = s_i'$ into equation 4.57:

$$t_f(UC) = \frac{c' \cot \phi' + s_i'}{\operatorname{cosec} \phi' - 1 + 2A_f} \tag{4.58}$$

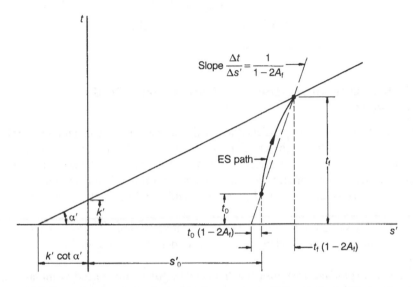

Figure 4.20 Example of undrained triaxial effective stress (ES) path for initial $K_0 \neq 1$.

EXAMPLE 4.5 UNDRAINED STRENGTH OF SOFT CLAY

Assuming a soft clay test specimen identical to those in Example 4.3, and $A_f = 0.8$, find the undrained compression strength c_u.

Solution

$$s_i' = -u_e = 32 \text{ kPa} \tag{Example 4.3}$$

Equation 4.58: $c_u = t_f(UC) = 10$ kPa

EXAMPLE 4.6 UNDRAINED STRENGTH OF STIFF CLAY

Assuming a stiff clay test specimen identical to those in Example 4.4 and $A_f = -0.15$, find the undrained compression strength c_u.

Solution

$$s_i' = -u_e = 107 \text{ kPa} \qquad \text{(Example 4.4)}$$

Equation 4.58: $c_u = t_f(\text{UC}) = 87 \text{ kPa}$

4.7 Relative short-term and long-term field strengths

The critical state concept predicts that a soil specimen under triaxial compression has an undrained strength which is determined by the water content and is independent of initial stress conditions and total stress path. The same is true for extension tests, but compression and extension values differ, as discussed in Section 2.4. Thus, the undrained strengths calculated for soft clay in Example 4.5 a n d stiff clay in Example 4.6 should also represent reasonable values to assume for relevant short-term field behaviour such as under foundation loading or in a cut slope.

Long-term drained strengths in the field will be greatly influenced by initial stress conditions and stress path, and it is impossible to reproduce these precisely in the laboratory. Thus, drained triaxial tests in the laboratory, while satisfactory for deriving values of c', ϕ', are of little use for predicting long-term strengths $t_f(\text{D1})$, $t_f(\text{D2})$. Better estimates of these may be made by inserting actual values of s_0', t_0 (as far as these are known) into equation 4.43 and equation 4.44. In a very approximate way, $t_f(\text{D1})$ can be thought of as the long-term strength under foundation loading and $t_f(\text{D2})$ can be thought of as the long-term strength in a cut slope.

EXAMPLE 4.7 SHORT-TERM AND LONG-TERM FIELD STRENGTHS OF SOFT CLAY

For the soft clay in Example 4.3, calculate the short-term strengths and approximate long-term strengths in the field under foundation loading and in a cut slope.

Solution

(a) Under foundation loading
Short-term strength $c_u = 10$ kPa (Example 4.5).
Long-term strength, $t_f(D1)$:

$$\sigma'_{v0} = 40 \text{ kPa}$$
$$\sigma'_{h0} = 28 \text{ kPa}$$
$$s'_0 = 34 \text{ kPa}$$
$$t_0 = 6 \text{ kPa}$$

Equation 4.43: $t_f(D1) = 17$ kPa

It can be seen that the long-term drained strength is much greater than the short-term strength, which will consequently govern except where the loading is applied very gradually. This behaviour is well known.

(b) In a cut slope
Short-term strength $c_u = 10$ kPa (Example 4.5).
Long-term strength $t_f(D2)$:

Equation 4.44: $t_f(D2) = 11$ kPa

As the short-term and long-term strengths differ very little, either may govern in the behaviour of the slope. In general it can be assumed that if the slope is stable when cut, it will remain stable, but show very little gain or loss of strength with time.

EXAMPLE 4.8 SHORT-TERM AND LONG-TERM FIELD STRENGTHS OF STIFF CLAY

For the stiff, heavily overconsolidated clay in Example 4.4, calculate the short-term strength and approximate long-term strength in the field under foundation loading and in a cut slope.

Solution

(a) Under foundation loading
Short-term strength $c_u = 87$ kPa (Example 4.6).
Long-term strength $t_f(D1)$:

$$\sigma'_{v0} = 70 \text{ kPa}$$
$$\sigma'_{h0} = 126 \text{ kPa}$$
$$s'_0 = 98 \text{ kPa}$$
$$t_0 = -56 \text{ kPa}$$

Equation 4.43: $t_f(\text{D1}) = 100 \text{ kPa}$

This indicates that while short-term strength will govern for design, the long-term strength will not necessarily be very much higher than the short-term strength.

(b) In a cut slope
Short-term strength $c_u = 87$ kPa (Example 4.6).
Long-term strength $t_f(\text{D2})$:

Equation 4.44: $t_f(\text{D2}) = 15 \text{ kPa}$

This indicates the well-documented behaviour that in a cut slope the strength of heavily overconsolidated clay can deteriorate very markedly with time. In a strongly fissured clay this deterioration can be very rapid. Strengths can also fall to considerably below the value of 15 kPa calculated above, in part because the value of $c' = 5$ kPa assumed in this calculation is high compared to values of 2 kPa or less that have been back-calculated from observed first time slips in clays of this type (e.g. Chandler and Skempton, 1974; Parry, 1988).

4.8 Infinite slope stress path

Long slopes may be treated as infinite in extent, allowing analysis to be made on the basis of a planar slip surface parallel to the slope itself. Such a surface is FF in Figure 4.21(a). Failure or creep can be brought about by an increase in pore pressure, and in examining the stability it is appropriate to duplicate this stress change in the triaxial cell or shear box, rather than performing more conventional tests.

Considering an element of slope ABCD in Figure 4.21(a), the weight W of a unit width is $\gamma h l \cos \beta$, where γ is the bulk unit weight of the soil. Making the usual assumption for an infinite slope, that the forces on faces AD and BC are equal and opposite, the shear stress along the potential slip surface DC and the total normal stress are

$$\text{Shear stress along DC} = \tfrac{1}{2}\gamma h \sin 2\beta$$

$$\text{Total normal stress along DC} = \gamma h \cos^2 \beta$$

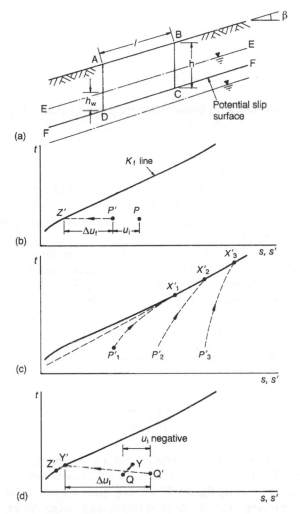

Figure 4.21 Infinite slope: (a) Section through slope; (b) field stress paths if soil weight constant; (c) conventional triaxial UI effective stress paths; (d) field stress paths if soil weight increases.

Assuming these to be t and s respectively, then

$$t = \tfrac{1}{2}\gamma h \sin 2\beta$$

$$s = \gamma h \cos^2 \beta$$

Assuming the piezometric line EE to be parallel to the slope and to be a height h_w above DC gives a pore pressure on DC equal to

$$u_i = \gamma h_w \cos^2 \beta$$

and thus

$$s' = \gamma(h - h_w) \cos^2 \beta$$

These total stress and effective stress states are given by points P, P′ in Figure 4.21(b). If the soil is already saturated at all depths, above and below the piezometric line, a rise in the piezometric line due, say, to heavy rainfall would not significantly alter t on plane DC, but would increase the pore pressure and decrease s'. The total stress point P would not change, but the effective stress would follow the path P′Z′ in Figure 4.21(b). If the slope were sufficiently steep and the rise in piezometric level sufficiently large, the stress path would reach the effective stress failure envelope (i.e. the K_f line) at Z′ and failure of the slope along FF in Figure 4.21a would occur.

The stress path P′Z′ does not correspond to any conventional drained path applied in the laboratory. It can be produced in the triaxial cell by applying σ_a, σ_r to the sample such that t and s at P, and t, s' at P′, are reproduced; then increasing the pore pressure (by back pressuring) keeping σ_a, σ_r constant until failure occurs.

For example, consider DC to be at a depth of 3 m, in a soil with $\gamma = 20$ kN/m³ and a surface slope of $\beta = 18°$. Assume h_w initially is 1 m and $\gamma_w = 10$ kN/m³.
Then initially

$$t = 17.6 \text{ kPa}$$
$$s = 54.1 \text{ kPa}$$
$$s' = 45.1 \text{ kPa}$$

If the piezometric line rose to the ground surface, t would remain the same, but s' would reduce to 27 kPa. Whether or not this would cause failure would depend on whether the point $t = 17.6$ kPa, $s' = 27$ kPa lay inside or outside the K_f line.

The level of stresses used above is realistic, and points to an aspect of considerable importance in this type of problem – the very low effective stress level. In making stability analyses for this type of problem the associated laboratory tests should be carried out at the correct stress level and with the appropriate stress paths. The use of such low stresses means that very accurate calibration of laboratory equipment is necessary, and sources of error such as ram friction in the triaxial test must be reduced to negligible levels.

A conventional procedure in triaxial soil testing is to carry out U1 undrained compression tests (with increasing axial stress) on a sample at its natural conditions, and then on two further samples consolidated under higher pressures. A heavily overconsolidated soil at its natural stress state initially will give an effective stress path such as P′₁X′₁ in Figure 4.21(b). A residual soil, which is usually compact, is also likely at this low stress level to produce a similar type of stress path. The stresses at the failure point X'_1 are much higher than at failure in the field and thus define a point on the K_f line remote from the field condition. The specimens consolidated under higher pressures will give effective stress paths such as $P'_2 X'_2$ and $P'_3 X'_3$. Thus X'_1, X'_2, X'_3 define a K_f line, but extrapolation of this line to the low stress region to define strength parameters is extremely hazardous. The K_f line in this region may be curved, and its location may also be strongly influenced by the actual stress path. Stress reduction can result in very strong dilation, which can influence the location of the strength envelope. Consequently, extrapolation of X'_1, X'_2, X'_3 to give effective stress strength parameters for use in analysing a potential shallow slip could be very conservative or very unsafe.

If the piezometric line is quite deep and lies below the potential slip surface, the stress conditions will be Q, Q' as shown in Figure 4.21(d). The initial pore pressure is negative. Under these conditions it is quite likely that the soil will be unsaturated for at least part of its depth, so that heavy rainfall, in saturating the ground, causes a rise in t as well as a drop in s', and an effective stress path such as Q'Y' will be followed. If the stress path reaches Y' failure will occur. A total stress path such as QY will be followed. Duplication of this change in a laboratory triaxial test would require the sample to be set up initially under field conditions, and then both the pore pressure and axial stress increased by appropriate amounts to produce failure.

Chapter 5

Elastic stress paths and small strains

5.1 Elastic behaviour in soils and soft rocks

Although soils and soft rocks do not generally behave in the linear and reversible manner of an elastic material, it is often found convenient, and sufficiently accurate, to assume elastic behaviour in the calculation of soil movements or stress changes in a soil mass arising from changes in boundary loading conditions. Immediate settlements of foundations on clay are commonly calculated using an elastic formula with an elastic modulus, obtained from undrained triaxial tests. The long-term settlements are calculated from deformation characteristics determined in consolidation tests, and stress changes in the soil below the foundation obtained from elastic formulae. Both the immediate and long-term settlements are based essentially on measured, large strain, deformation characteristics, which can be broadly distinguished as strains exceeding 0.1%. Increasing attention, however, is being paid to small strain behaviour, such as represented by the initial slopes of stress–strain curves. The assumption of elastic behaviour at these small strains is likely to be much more valid than for large strains.

5.2 Isotropic elastic stress paths

5.2.1 Elastic strains

Assuming an isotropic test specimen to be subjected to incremental principal effective stress increments $\Delta\sigma'_1$, $\Delta\sigma'_2$, $\Delta\sigma'_3$, the resulting strains ε_1, ε_2, ε_3, are given by:

$$\begin{vmatrix} \varepsilon_1 \\ \varepsilon_2 \\ \varepsilon_3 \end{vmatrix} = \frac{1}{E'} \begin{vmatrix} 1 & -v' & -v' \\ -v' & 1 & -v' \\ -v' & -v' & 1 \end{vmatrix} \begin{vmatrix} \Delta\sigma'_1 \\ \Delta\sigma'_2 \\ \Delta\sigma'_3 \end{vmatrix} \tag{5.1}$$

where E', $-v'$ are effective stress Young's modulus and Poisson's ratio respectively.

5.2.2 Undrained triaxial test

For a conventional triaxial compression test

$$\varepsilon_1 = \varepsilon_a \quad \Delta\sigma_1' = \Delta\sigma_a'$$

and

$$\varepsilon_2 = \varepsilon_3 = \varepsilon_r \quad \Delta\sigma_2' = \Delta\sigma_3' = \Delta\sigma_r'$$

Thus, from equation 5.1:

$$\varepsilon_a = \frac{1}{E'}(\Delta\sigma_a' - 2v'\Delta\sigma_r') \tag{5.2a}$$

$$\varepsilon_r = \frac{1}{E'}[\Delta\sigma_r' - v'(\Delta\sigma_a' + \Delta\sigma_r')] \tag{5.2b}$$

The volumetric strain ε_V is given by

$$\varepsilon_V = \varepsilon_a + 2\varepsilon_r \tag{5.3}$$

and thus,

$$\varepsilon_V = \frac{1}{E'}(1 - 2v')(\Delta\sigma_a' + 2\Delta\sigma_r') \tag{5.4}$$

But for undrained tests $\varepsilon_V = 0$ and as v' in general $\neq 0.5$,

$$\Delta\sigma_a' + 2\Delta\sigma_r' = 0 \tag{5.5}$$

i.e.

$$\Delta p' = 0 \tag{5.6}$$

which means that the mean effective stress is constant, giving a vertical effective stress path on a q–p' plot.

On a t–s' plot the stress path given by equation 5.5 is a straight line, the slope of which can easily be found by putting

$$\frac{\Delta t}{\Delta s} = \frac{\frac{1}{2}(\Delta\sigma_a' - \Delta\sigma_r')}{\frac{1}{2}(\Delta\sigma_a' + \Delta\sigma_r')} = 3.0 \tag{5.7}$$

This stress path is given in Figure 5.1 and, as shown, it extends into the extension zone. It can be seen that a specimen following an elastic stress path to failure will have different strengths in compression $t_f(C)$ and extension $t_f(E)$. The ratio of these strengths can be found from Figure 5.1 as follows:

$$\frac{t_f(C)}{s_0' + \frac{1}{3} t_f(C)} = \tan \alpha' \tag{5.8a}$$

$$\frac{t_f(E)}{s_0' - \frac{1}{3} t_f(E)} = \tan \alpha' \tag{5.8b}$$

From which

$$\frac{t_f(E)}{t_f(C)} = \frac{3 - \tan \alpha'}{3 + \tan \alpha'} \tag{5.9}$$

But, from equation 4.16, $\tan \alpha' = \sin \phi'$ and thus

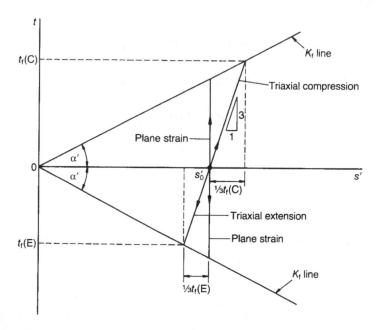

Figure 5.1 Effective stress paths for undrained tests on isotropic elastic soil or soft rock.

$$\frac{t_f(\text{E})}{t_f(\text{C})} = \frac{c_u(\text{E})}{c_u(\text{C})} = \frac{3-\sin\phi'}{3+\sin\phi'} \tag{5.10}$$

which is identical to equation 2.18, derived using the critical state concept, and which, as shown in Section 2.4, gives reasonable correlation with observed undrained strength ratios.

The correspondence between equations 5.10 and 2.18 arises because the critical state concept assumes that for initially identical specimens the mean effective stress p' at the critical state is the same in compression and extension. As seen in equation 5.6, $\Delta p' = 0$ for isotropic elastic behaviour, and thus initially identical specimens will have the same magnitudes of p' at failure.

5.2.3 Undrained plane strain

For plane strain it can be shown by putting $\varepsilon_2 = 0$ in equation 5.1 and

$$\varepsilon_V = \varepsilon_1 + \varepsilon_3 = 0$$

for undrained conditions that

$$\Delta\sigma_1 + \Delta\sigma_3 = 0 \tag{5.11}$$

The resulting stress path on a t vs s' plot is given by

$$\frac{\Delta t}{\Delta s'} = \frac{\frac{1}{2}\left(\Delta\sigma_1' - \Delta\sigma_3'\right)}{\frac{1}{2}\left(\Delta\sigma_1' + \Delta\sigma_3'\right)} \tag{5.12}$$

This gives a vertical stress path as shown in Figure 5.1.

The relationships between pore pressure change Δu and the change in total stress increments for isotropic elastic behaviour can be found by rewriting equations 5.5 and 5.11 in the following form

Equation 5.5 (triaxial): $(\Delta\sigma_a - \Delta u) + 2(\Delta\sigma_r - \Delta u) = 0$

from which

$$\Delta u = \Delta\sigma_3 + \tfrac{1}{3}\left(\Delta\sigma_a - \Delta\sigma_r\right) \tag{5.13}$$

Equation 5.11 (plane strain): $(\Delta\sigma_1 - \Delta u) + (\Delta\sigma_3 - \Delta u) = 0$

from which

$$\Delta u = \Delta\sigma_3 + \tfrac{1}{2}(\Delta\sigma_1 - \Delta\sigma_3) \qquad (5.14)$$

These equations can be compared with the Skempton (1948, 1954) expression:

$$\Delta u = \Delta\sigma_3 + A(\Delta\sigma_1 - \Delta\sigma_3) \qquad (5.15)$$

and thus $A = \tfrac{1}{3}$ for triaxial conditions and $\tfrac{1}{2}$ for plane strain.

EXAMPLE 5.1 COMPRESSION AND EXTENSION UNDRAINED STRENGTHS OF ISOTROPIC CLAY

Identical samples of reconstituted clay, prepared at a high moisture content, are consolidated isotropically in the triaxial cell under a pressure of 240 kPa. They are then allowed to swell to achieve an OCR = 3.0. If $c' = 0$, $\phi' = 23°$, what would be the undrained strengths of these specimens in compression and extension if they behave isotropically and elastically up to failure?

Solution

$$\text{Equation 4.16: } \tan\alpha' = \sin\phi'$$

$$\sigma_3' = \sigma_{cp} = \frac{240}{OCR} = 80\,\text{kPa}$$

From Figure 5.2,

$$t_f(C) = \left[80 + \tfrac{1}{3}t_f(C)\right]\tan\alpha'$$

$$= \left[80 + \tfrac{1}{3}t_f(C)\right]\sin 23°$$

$$\therefore \quad t_f(C) = c_u(C) = 35.9\,\text{kPa}$$

Similarly,

$$t_f(E) = c_u(E) = 27.6\,\text{kPa}$$

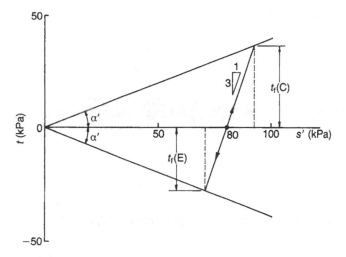

Figure 5.2 Example 5.1.

5.3 Undrained triaxial elastic stress paths in anisotropic soil or soft rock

Soils in the field are consolidated under anisotropic stress conditions giving them different stiffnesses in the vertical and horizontal directions. Soft normally consolidated or lightly overconsolidated clay will usually be stiffer in the vertical direction than the horizontal direction, whereas the opposite will hold for stiff heavily overconsolidated clays. It is useful then to see how anisotropy affects triaxial stress paths.

Consider a cross-anisotropic soil with an effective stress modulus E'_a in the axial or vertical direction, different from E'_r in the radial or horizontal direction. Poisson's ratio values are:

- v'_{ra} ratio of strain in the axial direction to an imposed strain in the radial direction;
- v'_{ar} ratio of strain in the radial direction to an imposed strain in the axial direction;
- v'_{rr} ratio of applied strain in one radial direction to an imposed strain in the orthogonal radial direction.

From superposition

$$\frac{v'_{ra}}{v'_{ar}} = \frac{E'_r}{E'_a} = n' \tag{5.16}$$

Linear strains ε_a, ε_r due to stress increments $\Delta\sigma'_a$, $\Delta\sigma'_r$ are given by

$$\varepsilon_a = \frac{\Delta\sigma'_a}{E'_a} - 2\frac{v'_{ra}\Delta\sigma'_r}{E'_r} \tag{5.17a}$$

$$\varepsilon_r = -v'_{ar}\frac{\Delta\sigma'_a}{E'_a} + \frac{\Delta\sigma'_r}{E'_r} - v'_{rr}\frac{\Delta\sigma'_r}{E'_r} \tag{5.17b}$$

Putting

$$\varepsilon_a + 2\varepsilon_r = 0$$

for undrained behaviour and combining equations 4.12, 4.13, 5.16, 5.17a and 5.17b gives

$$\frac{\Delta t}{\Delta s'} = \frac{2v'_{rr} + 4n'v'_{ar} - n' - 2}{2v'_{rr} + n' - 2} \tag{5.18}$$

and the Skempton pore pressure parameter A is given by

$$A = \frac{n'(1 - 2v'_{ar})}{n'(1 - 4v'_{ar}) + 2(1 - v'_{rr})} \tag{5.19}$$

Adoption of the assumption made by Henkel (1971) that

$$v'_{rr} = \tfrac{1}{2}(v'_{ar} + v'_{ra}) = \tfrac{1}{2}v'_{ar}(1 + n') \tag{5.20}$$

gives values of $\Delta t/\Delta s'$ and A as shown in Table 5.1 for n' ranging from 0.5 to 2.0, assuming $v'_{rr} = 0.2$.

Table 5.1 Values of $\Delta t/\Delta s'$ and Skempton pore pressure parameter A

n'	$\dfrac{\Delta t}{\Delta s'}$	A
0.5	1.4	0.15
0.8	2.1	0.27
1.0	3.0	0.33
1.6	∞	0.50
2.0	−6.3	0.58

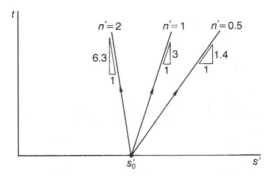

Figure 5.3 Effective stress paths for undrained triaxial compression tests on cross-anisotropic soil or soft rock.

It can be seen from this that anisotropy of a magnitude which might well occur in soil or soft rock can have a marked effect on stress path and pore pressure development. The stress paths for $n' = 0.5$, 1.0 and 2.0 are shown in Figure 5.3.

5.4 Observed effective stress paths for undrained triaxial tests

In many soils and soft rocks the early portion of the effective stress path in undrained tests agrees closely with that for isotropic or anisotropic elastic behaviour, until a yield condition or failure is approached. Observed effective stress paths are discussed below for a clay shale, a heavily overconsolidated clay, a moderately overconsolidated sandy clay and a soft recently deposited clay.

5.4.1 Clay shale

An experimental programme of tests carried out in the triaxial cell on Taylor clay shale at the US Army Waterways Experiment Station (Parry, 1976) included isotropic consolidation and swelling of a specimen and axial loading under undrained conditions.

Measurement of both volume change and axial strain during consolidation and swelling showed the specimen to be cross-anisotropic with a lateral stiffness higher than the vertical. In order to quantify the ratio n' of lateral stiffness E'_r to axial stiffness E'_a a value of v'_{rr} had to be assumed. Assuming $v'_{rr} = 0.2$ gave the following magnitudes of n' (where $n' = E'_r/E'_a$):

1. consolidation, $n' = 1.5$;
2. swelling, $n' = 1.7$.

The undrained triaxial test consisted firstly of a number of loadings up to about

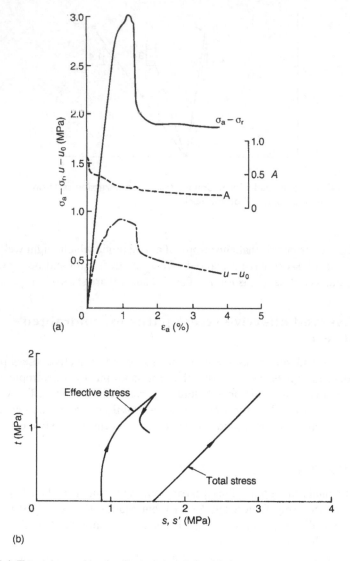

Figure 5.4 Triaxial test plots for Taylor clay shale: (a) deviator stress and pore pressure
against axial strain; (b) total and effective stress paths.

20% of the failure load, during which time the specimen showed linear and largely
reversible behaviour. It was then loaded to failure, which occurred at an axial strain
of 1.2%, and taken to a total axial strain of 4%. Plots of deviator stress $(\sigma_a - \sigma_r)$ pore
pressure change $(u - u_0)$, and pore pressure parameter A are shown plotted against
axial strain in Figure 5.4(a). The stress paths are shown in Figure 5.4(b). The value
of A measured in the initial loadings and the early part of the final loading was

$$A = 0.55 \text{ to } 0.60$$

Referring to Table 5.1, this suggests a lateral or radial stiffness approximately double the vertical or axial stiffness, which confirms the findings from the consolidation/swelling test of a higher lateral stiffness. The ratio $n' = 2$ indicated by the measured A values is higher than $n' = 1.5$ to 1.7 from the consolidation/swelling test, but this is of little significance as the strains measured in the latter test were extremely small and subject to substantial error. It is likely that $n' = 2.0$ is approximately correct.

It will be seen in Figure 5.4(b) that the early part of the effective stress path is almost vertical which reflects the high A value of around 0.6.

Subsequent tests on four other clay shales at WES (Leavell *et al.*, 1982) gave A values ranging from 0.52 to 0.70, indicating a ratio of n' of lateral stiffness to vertical stiffness of 1.6 to 2.6.

5.4.2 Heavily overconsolidated clay

A number of triaxial and plane strain tests were performed by Atkinson (1975) on undisturbed specimens of heavily overconsolidated London clay. He concluded that for small strain deformations the clay behaved as an anisotropic elastic material with the properties:

Vertical stiffness

$$E'_v = 11 \text{ MPa}$$
$$n' = 2.0$$
$$v'_{ar} = 0.2$$
$$v'_{rr} = 0$$

Putting $v'_{rr} = 0$ into equation 5.18, it can be seen that if $n' = 2$,

$$\frac{\Delta t}{\Delta s'} = \infty$$

and thus the stress path is vertical. Atkinson has shown stress paths in terms of σ'_a and σ'_r, for which he gives the stress path gradient m, where

$$m = \frac{\Delta \sigma'_a}{\Delta \sigma'_r}$$

A value of $m = -1$ corresponds to

$$\frac{\Delta t}{\Delta s'} = \infty$$

Atkinson also lists m values for London clay obtained by a number of other

Table 5.2 Values of $\Delta t/\Delta s'$ and n' for London clay

m	$\dfrac{\Delta t}{\Delta s'}$	n'
−0.7	−5.67	2.4
1.0	∞	2.0
−1.24	+9.33	1.75

workers and these range, for triaxial tests on vertical specimens, from $m = 0.7$ to $m = 1.24$; these values appear to be the same for compression tests and extension tests, and thus independent of whether the effective stresses are increasing or decreasing. Taking $v'_{\mathrm{rr}} = 0$, $v'_{\mathrm{ar}} = 0.2$ gives the values of n' in Table 5.2.

5.4.3 Moderately overconsolidated sandy clay

The triaxial test described here was carried out on a specimen of very stiff sandy silty clay, taken from a depth of 18.2 m below sea bed, at the site of the Heather Production Platform in the northern North Sea (Parry, 1980, tests performed by Fugro Limited). Based on the liquidity index (NAVFAC, 1971) and undrained shear strength (Vijayvergia, 1977) it was estimated that, at this depth, OCR = 9 approximately. Index properties at this depth were $w_{\mathrm{L}} = 35\%$, $w_{\mathrm{P}} = 15\%$. It was found from a number of tests that effective stress strength parameters could be taken as $c' = 0$, $\phi' = 33°$ for design purposes.

The test specimen from 18.2 m depth was initially placed in the triaxial cell under a cell pressure of 680 kPa and back pressure 300 kPa. It was then submitted to an undrained compression test, for which the plots of deviator stress and pore pressure against axial strain are shown in Figure 5.5(a), and the resulting stress paths in Figure 5.5(b). It will be noted that the stress path is linear, with a slope of 2.1:1, until it reaches an envelope through the origin giving maximum stress obliquity ($\alpha' = 30°$, $\phi' = 35°$), after which it first follows this envelope before falling away slightly from it until the maximum t' is reached. Referring to Table 5.1 this indicates $n' = 0.8$; that is, a slightly greater stiffness vertically than horizontally.

5.4.4 Soft recently deposited clay

It has been shown (Parry and Wroth, 1981) for a number of soft recently deposited clays, by conducting consolidation tests on vertical and horizontal specimens, that the vertical stiffness is about twice the horizontal stiffness. This suggests a value of $n' = 0.5$, and, referring to Table 5.1, an undrained triaxial test on a vertical specimen would be expected to give a stress path with an initial slope of about 1.4:1 with a corresponding $A = 0.15$. In fact some soft clays behave more like

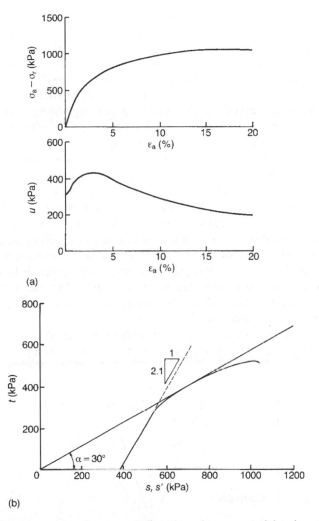

Figure 5.5 Triaxial test plots for a very stiff moderately overconsolidated sandy clay:
(a) deviator stress and pore pressure against axial strain; (b) total and
effective stress paths.

isotropic elastic materials. This is because the axial effective stress σ'_a increases, but the radial effective stress σ'_r decreases because of the pore pressure increase. In soft clays the stiffness moduli may be different for increasing and decreasing effective stress and, in general, the value for decreasing stress may be about double that for increasing stress in any specific direction. This compensates for the different vertical and horizontal stiffnesses measured in consolidation tests under increasing stress, and results in the vertical specimens behaving in a more or less isotropic manner (Parry and Wroth, 1981).

If a horizontal specimen is tested in undrained triaxial compression the anisotropy is exaggerated by the differing moduli for increasing and decreasing effective stress. A horizontal specimen will have the vertical field stiffness in one radial direction and the horizontal stiffness in the other. For purposes of illustration here it will be assumed that the radial stiffness is that given by the mean of the two. While not strictly correct, this should give a reasonable prediction for stress path studies. This means that if the stiffness under decreasing effective stress is again assumed to be double that for increasing effective stress then, effectively, $n' = 3.0$ during an undrained triaxial test on a horizontal specimen.

In Figure 5.6 the broken lines are the elastic stress paths assuming a vertical stiffness twice the horizontal stiffness and not influenced by whether the stresses are increasing or decreasing. That is, $n' = 0.5$ for the vertical specimen and $n' = 1.5$ for the horizontal specimen. The full lines are the corresponding elastic stress paths, assuming the modulus for decreasing stress is twice that for increasing stress. It can be seen that the stress paths have rotated anticlockwise, giving expected stress paths OB for a vertical test specimen and OD for a horizontal specimen. Assuming $v'_{rr} = 0.2$ the gradients of the stress paths in Figure 5.6 are given in Table 5.3.

Similar reasoning to that above can be applied to extension tests on vertical and horizontal specimens. The resulting stress paths are shown in Figure 5.6 and have the gradients given in Table 5.4.

There is little published evidence of undrained triaxial compression and extension tests with pore pressure measurement on natural soft clay samples, taken

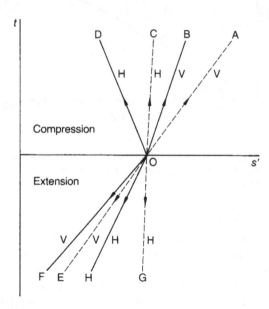

Figure 5.6 An example of the possible influence of sample orientation on effective stress paths in cross-anisotropic soil.

Table 5.3 Compression tests

Orientation	n'	Stress path	Gradient
Vertical	0.5	OA	1.4
Vertical	1.0	OB	3.0
Horizontal	1.5	OC	21
Horizontal	3.0	OD	−2.4

Table 5.4 Extension tests

Orientation	n'	Stress path	Gradient
Vertical	0.5	OE	1.4
Vertical	0.25	OF	1.1
Horizontal	1.5	OG	21
Horizontal	0.75	OH	2.0

vertically and horizontally. One reason for this may be the need to test at low effective stress levels consistent with field stresses. A notable exception is the work of Wesley (1975) on soft clay from Mucking Flats adjacent to the River Thames. The water table at this site is about 1 m below ground level, and other relevant data here are for the soil between depths of 3.18 m and 3.48 m, for which $w_L - 56\%$, $w_p = 25\%$, $w = 52\%$, the sensitivity from vane tests is about 5, and the measured overconsolidation ratio is about 1.5. The clay fraction is about 30%.

Undrained triaxial compression and extension tests were carried out by Wesley on horizontal and vertical undisturbed samples first consolidated to the assumed mean effective stress in the field, equal to 20.7 kPa. One-dimensional consolidation tests on vertical and horizontal specimens gave, for small stress changes above field values, a horizontal compressibility almost twice the vertical compressibility. The stress paths for the undrained tests are shown on a t–s' plot in Figure 5.7 with the predicted elastic stress paths (broken lines) superimposed.

Although the observed stress paths do not show initial linearity, there is nevertheless an excellent correspondence between the patterns of observed and predicted paths. This suggests, at least for this soft clay, that the horizontal stiffness is about one-half the vertical stiffness for increasing effective stresses, and that the stiffness for decreasing effective stress is about twice that for increasing effective stress.

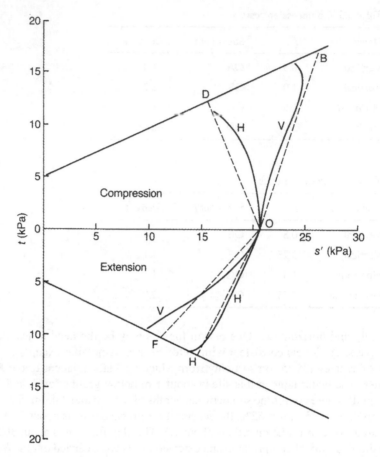

Figure 5.7 Observed stress paths for undrained triaxial tests on vertical and horizontal specimens of soft Mucking Flats clay. (After Wesley, 1975.)

EXAMPLE 5.2 PORE PRESSURE AT FAILURE IN ANISOTROPIC CLAY

A sample of stiff clay with effective stress strength characteristics $c' = 0$, $\phi' = 24°$ is initially at equilibrium in the triaxial cell under a cell pressure of 200 kPa and zero pore pressure. It is taken to failure in undrained triaxial compression, and the measured undrained shear strength is 77 kPa. Assuming the soil behaves elastically up to failure with respect to stress changes, estimate the value of the anisotropy parameter n' if $v'_{rr} = 0.2$. What is the pore pressure value at failure and the pore pressure parameter A_f?

Solution

$$\text{Equation 4.16: } \tan \alpha' = \sin 24°$$
$$\therefore \quad \alpha' = 22.1°$$
$$s_\mathrm{f}' = 77 \cot \alpha' = 190 \text{ kPa}$$

Referring to Figure 5.8, the slope of effective stress path AB is

$$-\frac{77}{200-190} = -7.7$$

Substituting

$$\frac{\Delta t}{\Delta s'} = -7.7$$

into equation 5.18:

$$-7.7 = \frac{2v_\mathrm{rr}' + 4n' v_\mathrm{ar}' - n' - 2}{2v_\mathrm{rr}' + n' - 2}$$

Equation 5.20:

$$v_\mathrm{ar}' = \frac{2v_\mathrm{rr}'}{1+n'} = \frac{0.4}{1+n'}$$

Substituting v_ar' and v_rr' into equation 5.18 gives

$$n' = 1.9$$

Pore pressure at failure u_f, from Figure 5.8:

$$u_\mathrm{f} = 77 + 10 = 87 \text{ kPa}$$

Equation 5.14, putting $\Delta\sigma_3 = 0$:

$$A_\mathrm{f} = \frac{u_\mathrm{f}}{\Delta\sigma_1} = \frac{87}{2 \times 77} = 0.56$$

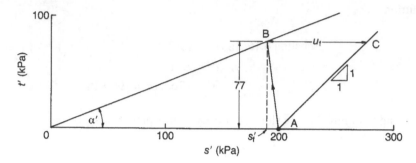

Figure 5.8 Example 5.2.

EXAMPLE 5.3 ANISOTROPY AND PORE PRESSURE CHANGE IN SOFT ROCK

A sample of clay shale is set up in the triaxial cell under stress conditions thought to exist in the field, namely a total axial stress of 1000 kPa, a total radial stress of 2000 kPa and a pore pressure (back pressure) of 300 kPa. On increasing the total axial stress to 2000 kPa, keeping the total radial stress constant, it is observed that the pore pressure increases to 900 kPa. If the rock is behaving elastically and $v'_{rr} = 0.2$, calculate the value of the anisotropy parameter n'.

Solution

$$\Delta \sigma_1 = 1000 \text{ kPa}$$
$$\Delta \sigma_3 = 0$$
$$\Delta u = 600 \text{ kPa}$$

Equation 5.14: $A = 0.6$

Equation 5.20: $v'_{ar} = \dfrac{0.4}{1+n'}$

Equation 5.19: $A = \dfrac{n'(1-2v'_{ar})}{n'(1-4v'_{ar})+2(1-v'_{rr})}$

From which it can be found that

$$n' = 2.1$$

5.5 Small strain behaviour

The increasing emphasis on small strain behaviour has arisen because:

(a) under working stresses in the field strains are often quite small and attempting to interpolate these from large strains measured in the laboratory can lead to large errors;
(b) small strains arising from geotechnical activities, such as tunnelling or deep excavation, in urban areas can affect existing structures and it is essential to calculate movements with the greatest possible accuracy;
(c) sophisticated methods are now available for measuring small strains, both in the laboratory and in the field;
(d) improved constitutive models and computer based calculations enable these measured parameters to be applied effectively.

The strain ranges capable of being measured by different laboratory techniques and typical strain ranges for various structures are shown in Figure 5.9 (Atkinson, 2000).

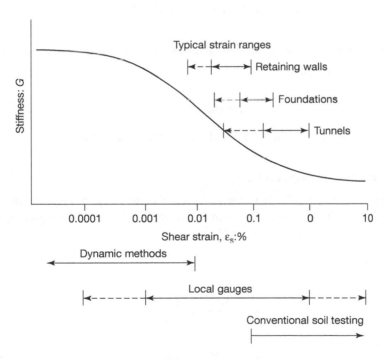

Figure 5.9 Strain ranges capable of being measured by different laboratory techniques and typical strain ranges for various structures (Atkinson, 2000). Reproduced by kind permission of Thomas Telford.

5.6 Elastic small strain behaviour in isotropic geomaterials

5.6.1 Shear modulus

Under undrained conditions, the value of shear modulus G^u in terms of total stress has the same magnitude as the effective stress value G′. This equality arises because the pore water offers no resistance to shear.

Consider a conventional undrained triaxial compression test on a saturated specimen submitted to an applied axial total stress increment $\Delta\sigma_a$ with radial total stress σ_r constant, leading to axial and radial strains ε_a, ε_r respectively.

$$G^u = \frac{\tau}{\gamma} = \frac{\Delta\sigma_a - \Delta\sigma_r}{2\gamma} \tag{5.21a}$$

$$G' = \frac{\tau}{\gamma} = \frac{\Delta\sigma_a' - \Delta\sigma_r'}{2\gamma} \tag{5.21b}$$

As the deviator stress is identical in terms of total and effective stresses:

$$G^u = G' = G \tag{5.22}$$

Shear strain γ is given by:

$$\gamma = (\varepsilon_a - \varepsilon_r) \tag{5.23}$$

but, as ε_r, ε_a are opposite in direction, their magnitudes are additive.

5.6.2 Elastic parameters

Denoting total stress and effective stress Poisson's ratios as γ^u, γ' respectively, the shear modulus for an isotropic soil can be evaluated in terms of these parameters and the respective Young's moduli E^u, E′.

Total stress

$$\varepsilon_a = \frac{\Delta\sigma_a}{E^u} - 2v^u \frac{\Delta\sigma_r}{E^u} \tag{5.24a}$$

or

$$\varepsilon_a = \frac{\Delta\sigma_a}{E^u} \text{ as } \Delta\sigma_r = 0 \tag{5.24b}$$

from which E^u can be evaluated. Similarly:

$$\varepsilon_r = -v^u \frac{\Delta\sigma_a}{E^u} \tag{5.24c}$$

The negative sign is required because Poisson's ratio is expressed as a positive number, whereas ε_a, ε_r are opposite in direction.

Combining equations 5.21a, 5.23, 5.24b and 5.24c, and putting $\Delta\sigma_r = 0$ gives:

$$G = \frac{E^u}{2(1+v^u)} \tag{5.25}$$

In an undrained test on a saturated specimen the volume change ε_v is zero:

$$\varepsilon_V = \varepsilon_1 + \varepsilon_2 + \varepsilon_3 = 0 \tag{5.26}$$

but, for triaxial compression, $\varepsilon_1 = \varepsilon_a$, $\varepsilon_2 = \varepsilon_3 = \varepsilon_r$ and thus, from equation (5.26), $\varepsilon_r = -0.5\varepsilon_a$, giving $v^u = 0.5$, and:

$$G = \frac{E^u}{3} \tag{5.27}$$

Effective stress

$$\varepsilon_a = \frac{1}{E'}(\Delta\sigma'_a - 2v'\Delta\sigma'_r) \tag{5.28a}$$

$$\varepsilon_r = \frac{1}{E'}(-v'\Delta\sigma'_a - v'\Delta\sigma'_r + \Delta\sigma'_r) \tag{5.28b}$$

Combining equations 5.21b, 5.23, 5.28a and 5.28b gives:

$$G = \frac{E'}{2(1+v')} \tag{5.29}$$

5.7 Elastic small strain behaviour in cross-anisotropic geomaterials

Many deposited soils and some soft rocks, such as mudstones and clay-shales, are horizontally bedded, or very nearly so, and exhibit a horizontal stiffness uniform in all directions, but differing from the vertical stiffness. This is known as cross-anisotropy. In soft, recently deposited clays and silts, the horizontal stiffness may be 0.5 to 0.8 times the vertical value, while in heavily over-consolidated soils and soft rocks, from which large depths of overburden have been removed, the horizontal stiffness may typically be 2 to 3 times the vertical value.

5.7.1 Cross-anisotropic elastic parameters

A cross-anisotropic elastic material is completely defined by seven parameters:

E_v = Young's modulus in the vertical direction
E_h = Young's modulus in the horizontal direction
v_{vh} = ratio of horizontal strain to an imposed axial strain
v_{hv} = ratio of axial strain to an imposed horizontal strain
v_{hh} = ratio of horizontal strain to an imposed strain in the orthogonal horizontal direction
G_{hv} = shear modulus in the vertical plane $(= G_{vh})$
G_{hh} = shear modulus in the horizontal plane.

Cartesian coordinate z corresponds to the v direction and x, y to orthogonal h directions. For a triaxial specimen sampled vertically, suffixes a, r are commonly used and correspond to v, h respectively.

Not all the above are independent parameters. As the cross-anisotropic material is isotropic in the horizontal direction, it follows from equation 5.9 that:

$$G_{hh} = \frac{E_h}{2(1+v_{hh})} \tag{5.30}$$

There is also, for an elastic material, a thermodynamic requirement that:

$$\frac{v_{hv}}{E_h} = \frac{v_{vh}}{E_v} \tag{5.31}$$

A full description of a cross-anisotropic elastic soil or rock therefore requires evaluation of five independent parameters, usually taken to be E_v, E_h, v_{vh}, v_{hh}, G_{hv} as shown in equation 5.12:

$$
\begin{vmatrix} \varepsilon_x \\ \varepsilon_y \\ \varepsilon_z \\ \gamma_{yz} \\ \gamma_{zx} \\ \gamma_{xy} \end{vmatrix}
=
\begin{vmatrix}
1/E_h & -v_{hh}/E_h & -v_{vh}/E_v & & & \\
-v_{hh}/E_h & 1/E_h & -v_{vh}/E_v & & & \\
-v_{vh}/E_v & -v_{vh}/E_v & 1/E_v & & & \\
& & & 1/G_{hv} & & \\
& & & & 1/G_{hv} & \\
& & & & & 2(1+v_{hh})/E_h
\end{vmatrix}
\begin{vmatrix} \Delta\sigma_x \\ \Delta\sigma_y \\ \Delta\sigma_z \\ \Delta\tau_{yz} \\ \Delta\tau_{zx} \\ \Delta\tau_{xy} \end{vmatrix} \tag{5.32}
$$

Undrained total stress

In an undrained compression test on an axisymmetric triaxial specimen, the zero volume change imposes the conditions:

$$v_{ar}^u = v_{vh}^u = \tfrac{1}{2} \tag{5.33a}$$

$$v_{ra}^u + v_{rr}^u = v_{hv}^u + v_{hh}^u = 1 \tag{5.33b}$$

Expressing the ratio of horizontal to vertical moduli as n^u:

$$n^u - \frac{E_r^u}{E_a^u} \tag{5.34}$$

From equations 5.31 and 5.34:

$$n^u = \frac{v_{ra}^u}{v_{ra}^u} \tag{5.35}$$

From equations 5.33a, 5.33b and 5.35:

$$v_{ra}^u = 0.5 n^u \tag{5.36a}$$

$$v_{rr}^u = (1 - 0.5 n^u) \tag{5.36b}$$

From equations 5.30 and 5.36b:

$$E_r^u = 2G_{rr}(1 + v_{rr}^u) = 2G_{rr}(2 - 0.5 n^u) \tag{5.37}$$

For very small strains the value of $G_{rr} = G_{0rr}$ can be evaluated by dynamic tests in the laboratory or the field, passing shear (S) waves through the soil or rock. In the field shear waves generated at a chosen depth in a borehole are detected at the same depth by geophones in a line of boreholes aligned to the source, or by pushed-in probes. The times are normally recorded between receivers, rather than from the source. In the laboratory, bender elements (Shirley and Hampton, 1978) are now commonly used, made from piezoelectric material. These can be inserted into the pedestal and top cap of a triaxial test as shown in Figure 5.10, such that the strips penetrate about 3 mm into the test specimen. By electrically exciting one element, S waves are passed through the specimen, generating a detectable voltage in the other element. The time taken for the wave to pass through the specimen can be converted into a shear wave velocity V_S, from which the shear modulus, denoted by G_0 (or G_{max}), can be obtained from:

$$G_0 = \rho V_{s2} \tag{5.38}$$

for an isotropic soil, where ρ is bulk soil density.

In the triaxial cell, bender elements in the pedestal and top cap as shown in Figure 5.10 will measure V_{Sar} ($= V_{Svh}$) in an axial symmetric anisotropic specimen. V_{Sra} can be measured by bender elements attached to the perimeter of the specimen. The value of V_{Srr} ($= V_{Shh}$) can be obtained either by attaching bender elements to the perimeter of the specimen or by having bender elements in the pedestal and top cap, carefully oriented to give this value for a specimen sampled horizontally.

Insertion of a dynamically determined value of G_{0rr} into equation 5.37 still requires a value of n_0^u to evaluate E_{0r}^u. A possible method (Atkinson, 2000) is to make direct physical measurements of strains in the triaxial test and to assume that the parameters deduced in this way also hold for the much smaller dynamic strains. Such strains should be measured precisely by sensitive devices attached directly to, or lightly contacting, the triaxial test specimen.

In carefully conducted undrained triaxial tests in the laboratory on specimens initially consolidated under assumed field stresses, with rest periods during loading to counter creep effects, Clayton and Heymann (2001) found for three very different geomaterials, soft Bothkennar clay, stiff London clay and chalk, linear behaviour up to axial strains of 0.002% to 0.003%. Typical elastic moduli E_{max} were 24 MPa, 240 MPa and 4800 MPa respectively. The ratio of E_{max} in the laboratory to E_0 obtained by geophysical methods in the field was close to unity for the soft clay, 0.75 for the stiff clay, possibly as a result of some sample

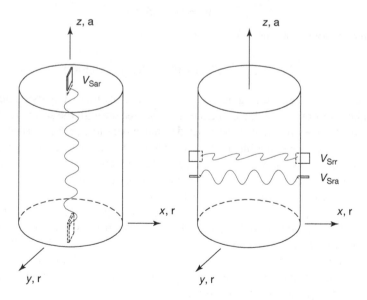

Figure 5.10 Application of bender elements to triaxial samples.

disturbance, and about 4 for the chalk, probably reflecting the influence of fissuring on measured surface wave velocities.

Effective stress

The ratio of radial to vertical effective stress moduli for an axisymmetric triaxial specimen is denoted n', thus:

$$n' = \frac{E'_r}{E'_a} \tag{5.39}$$

And also, to satisfy thermodynamic requirements:

$$n' = \frac{v'_{ra}}{v'_{ar}} \tag{5.40}$$

The five small strain effective stress parameters E'_{0a}, E'_{0r}, v'_{0ar}, v'_{0rr}, G_{0ra} governing axisymmetric behaviour can be obtained from a combination of dynamic bender tests and triaxial tests with selected stress paths. Small strain drained triaxial tests holding either axial effective stress σ'_a constant or radial effective stress σ'_h constant yield the parameters E'_{0a}, E'_{0r}, v'_{0ar} $(1 - v'_{0rr})/E'_{0r}$.
From equation 5.32, for small strains:

$$\begin{vmatrix} \varepsilon_a \\ \varepsilon_r \end{vmatrix} = \begin{vmatrix} 1/E'_{0a} & -2v'_{0ra}/E'_{0r} \\ -v'_{0ar}/E'_{0a} & (1 - v'_{0rr})/E'_{0r} \end{vmatrix} \begin{vmatrix} \Delta\sigma'_a \\ \Delta\sigma'_r \end{vmatrix} \tag{5.41}$$

Test holding σ'_r constant

If an axial compression stress increment $\Delta\sigma'_a$ is applied to a cross-anisotropic test specimen while holding the radial effective stress σ'_r constant then, from equation 5.41:

$$\varepsilon_a = \frac{\Delta\sigma'_a}{E'_{0a}} \tag{5.42a}$$

$$\varepsilon_r = \frac{-v'_{0ar}\,\Delta\sigma'_a}{E'_{0a}} \tag{5.42b}$$

Equation 5.42a yields the value of E'_{0a} from the measured value of axial strain ε_a. Substituting this value of E'_{0a} into equation 5.42b allows v'_{0ar} to be evaluated.

Test holding σ'_a constant

If a radial compression stress increment $\Delta\sigma'_r$ is applied to a cross-anisotropic test specimen while holding the axial effective stress $\Delta\sigma'_a$ constant then, from equation 5.41:

$$\varepsilon_a = -2v'_{0ra}\frac{\Delta\sigma'_r}{E'_{0r}} \tag{5.43a}$$

$$\varepsilon_r = \frac{\Delta\sigma'_r}{E'_{0r}}(1-v'_{0rr}) \tag{5.43b}$$

Equations 5.43a and 5.43b both have two unknowns and these can be resolved by the use of bender tests. In terms of the triaxial test specimen, equation 5.30 becomes:

$$G_{0rr} = \frac{E'_{0r}}{2(1+v'_{0rr})} \tag{5.44}$$

Substituting the measured values of ε_r and G_{0rr} into equations 5.43b and 5.44 respectively yields the values of E'_r and v'_{rr}.
  Putting

$$\Omega = \frac{2G_{0rr}\varepsilon_r}{\Delta\sigma'_r} \tag{5.45}$$

from equations 5.43b and 5.44:

$$v'_{0rr} = \frac{1-\Omega}{1+\Omega} \tag{5.46}$$

and

$$E'_{0r} = \frac{4G_{0rr}}{1+\Omega} \tag{5.47}$$

The value of v'_{0ra} can be found by substituting E'_{0r} from equation 5.47 together with the measured value of axial strain ε_a into equation 5.43a. For an ideal cross-anisotropic material this should have the same value as v'_{0ar}.

EXAMPLE 5.4 SMALL STRAIN PARAMETERS

A vertical sample taken from a depth of 10 m in a stiff, heavily overconsolidated deposit of clay is to be tested in the triaxial cell to determine its small strain characteristics. The unit weight of the soil is 20 kN/m³ and the ground water conditions are at equilibrium with a water table at 2 m depth. Assume $K_{-} = 2.0$ and unit weight of water is 10 kN/m³. In preparation for small strain tests the specimen is initially set up in the triaxial cell with effective stress conditions identical to those believed to exist in the field. Assume the soil during the tests behaves as a fully reversible, linear elastic material.

Part A
A drained axial compression test is performed holding the cell pressure constant and increasing the axial stress by an increment equal to one-half the *in situ* vertical total stress, giving measured axial and lateral strains respectively of 0.080% and –0.002%. After removing the axial stress increment a second drained test is performed holding the axial stress constant and increasing the radial stress by an amount equal to one-half the *in situ* lateral total stress, giving axial and radial strains respectively of –0.006% and 0.050%. A cross-borehole dynamic test in the field has indicated $G_{0hh} = 144\,000$ kPa. Determine E'_{0v}, E'_{0h}, v'_{0vh}, v'_{0hh}. Compare n' values obtained from both relative modulus values and relative Poisson's ratio values, and comment.

Part B
After removing the lateral stress increment an undrained test is performed, keeping the radial total stress constant and increasing the axial total stress by an amount equal to the *in situ* vertical total stress, giving an axial strain of 0.064%. Calculate E^u_{0a} and E^u_{0r}.

Solution

Field stresses
$$\sigma_v = 10 \times 20 = 200 \text{ kPa}$$
$$\sigma'_v = 200 - 8 \times 10 = 120 \text{ kPa}$$
$$\sigma'_h = 120 \times 2 = 240 \text{ kPa}$$
$$\sigma_h = 240 + 80 = 320 \text{ kPa}$$

First drained test
$$\Delta\sigma'_a = 100 \text{ kPa} \quad \varepsilon_a = 0.08\% \quad \varepsilon_r = -0.002\%$$

Equation 5.42a: $E'_{0v} = E'_{0a} = \dfrac{\Delta\sigma'_a}{\varepsilon_a} = \dfrac{100 \times 100}{0.08} = 125\,000 \text{ kPa}$

Equation 5.42b: $v'_{0vh} = v'_{0ar} = -\dfrac{\varepsilon_r E'_{0a}}{\Delta \sigma'_a} = -\dfrac{-0.002 \times 125\,000}{100 \times 100} = 0.025$

Second drained test + dynamic field test

$$G_{0rr} = G_{0hh} = 144\,000 \text{ kPa}$$

Equation 5.45: $\Omega = \dfrac{2G_{rr}\varepsilon_r}{\Delta \sigma'_r} = \dfrac{2 \times 144\,000 \times 0.050}{100 \times 100} = 0.90$

Equation 5.46: $v'_{0rr} = \dfrac{1-\Omega}{1+\Omega} = \dfrac{1-0.90}{1+0.90} = 0.052$

Equation 5.47: $E'_{0r} = \dfrac{4G_{rr}}{1+\Omega} = \dfrac{4 \times 144\,000}{1.90} = 303\,000 \text{ kPa}$

Equation 5.43a: $v'_{0ra} = -\dfrac{\varepsilon_a E'_{0r}}{2\Delta \sigma'_r} = -\dfrac{-0.006 \times 303\,000}{2 \times 100 \times 160} = 0.057$

Equation 5.40: $n'_0 = \dfrac{v'_{0ra}}{v'_{0ar}} = \dfrac{0.057}{0.025} = 2.28$

Equation 5.39: $n'_0 = \dfrac{E'_{0r}}{E'_{0a}} = \dfrac{303\,000}{125\,000} = 2.42$

In an ideal cross-anisotropic medium these n'_0 values would be equal, but in fact the agreement here is remarkably good in view of the very small measured strains on which they are based.

Undrained test

$$E^u_{0a} = \dfrac{\Delta \sigma_a}{\varepsilon_a} = \dfrac{200 \times 100}{0.064} = 312\,500 \text{ kPa}$$

Equation 5.34: $E^u_{0r} = n^u_0 E^u_{0a} = 312\,500\, n^u_0 \text{ kPa}$

Equation 5.37: $E^u_{0r} = 2G_{0rr}\,(2 - 0.5n^u_0)$

Thus $312\,500 n^u_0 = 2 \times 144\,000(2 - 0.5n^u_0)$

from which $n_0^u = 3.42$

giving $E_{0r}^u = 1\,069\,000$ kPa

Chapter 6

The use of stress discontinuities in undrained plasticity calculations

6.1 Lower bound undrained solutions

If a saturated soil at failure is assumed to be deforming plastically at constant volume (i.e. without any dissipation of pore pressure), simple solutions to stability problems can often be obtained by ensuring that, at any point within the soil mass, **total stress equilibrium** is satisfied and the **total stress failure criterion** is not violated. These solutions give a **lower bound** on the applied loading.

In most problems the geometry is such that it is necessary to assume one or more changes in the stress field to satisfy boundary conditions. This can be accomplished by inserting total stress discontinuities within the deforming mass.

6.2 Smooth retaining wall

A simple problem which can be solved without having to assume a stress discontinuity is the case of soil retained by a smooth retaining wall as shown in Figure 6.1. The solutions to obtain the active and passive pressures on the wall are shown in Figures 6.1(a) and 6.1(b) respectively. It is assumed that the vertical stress at all points in the soil behind the wall is a principal stress, equal to the overburden pressure, i.e.

$$\sigma_v = \gamma z \tag{6.1}$$

where z is the depth below ground level and γ is the unit weight of soil, assumed to be constant.

It can be seen in Figure 6.1 that the horizontal stress, and consequently the pressure on the wall, is given by

$$\text{Active pressure} \quad \sigma_h = \sigma_v - 2c_u \tag{6.2a}$$

$$\text{Passive pressure} \quad \sigma_h = \sigma_v + 2c_u \tag{6.2b}$$

Substituting equation 6.1 into equations 6.2a and 6.2b gives

$$\text{Active pressure} \quad \sigma_h = \gamma z - 2c_u \qquad (6.3\text{a})$$

$$\text{Passive pressure} \quad \sigma_h = \gamma z + 2c_u \qquad (6.3\text{b})$$

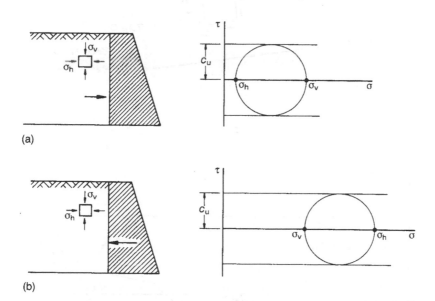

(a)

(b)

Figure 6.1 Undrained total stress circles for a smooth retaining wall: (a) active case; (b) passive case.

Equation 6.3a gives a tensile lateral stress for depths z less than $2c_u/\gamma$. Although this is a mathematically valid solution, it is unrealistic to assume that such a tensile stress could be maintained between the soil and the wall, and the total stress method in consequence is not satisfactory for calculating active pressure generated by soil behind a retaining wall. An approach sometimes adopted is to assume tension cracks open up to a depth of $2c_u/\gamma$ and use equation 6.3a to calculate lateral pressures below this depth, assuming a surcharge of γz to act above this level. This raises the further problem that water may fill the crack and exert pressure on the wall. A designer will wish to add this pressure to the calculation, which gives a hybrid result as water pressure should not need to be calculated separately in a total stress solution.

6.3 Stress discontinuity

In most problems the geometry is such that it is necessary to assume one or more changes in the stress field in order to satisfy boundary conditions. This can be accomplished by inserting stress discontinuities within the deforming mass. Figure 6.2(a) shows a linear stress discontinuity separating different stress

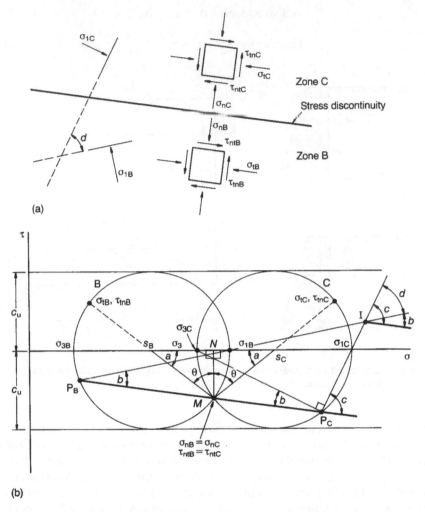

Figure 6.2 Total stress discontinuity: (a) stresses in zones B and C; (b) stress circles.

zones B* and C in a soil mass deforming under undrained plane strain conditions. In order that stress equilibrium is satisfied, it is necessary that the normal total stresses and shear stresses should equate across the discontinuity, i.e.

$$\sigma_{nC} = \sigma_{nB} \qquad (6.4a)$$

---

* The letter A has not been used in order to avoid possible confusion with its use to denote an active zone.

$$\tau_{ntC} = \tau_{ntB} \tag{6.4b}$$

but equality of the direct stresses in the direction of the discontinuity is not necessary, i.e.

$$\sigma_{tC} \neq \sigma_{tB} \tag{6.4c}$$

which allows a change in stress field across the discontinuity.

The stress conditions on either side of the discontinuity shown in Figure 6.2(a) are given by the stress circles B and C in Figure 6.2(b). The geometry of Figure 6.2(b) satisfies the requirements of equations 6.4a and 6.4b, with the lower intersection point M of the two circles representing these stresses, as the shear stresses τ_{ntB} and τ_{ntC} are clockwise in Figure 6.2(a) and are thus negative.

A line drawn through the intersection point M parallel to the discontinuity establishes the pole points for planes P_B and P_C. The directions of the planes on which the major principal stresses act in the two zones are given by lines drawn through $P_B \sigma_{1B}$ for zone B and $P_C \sigma_{1C}$ for zone C. Denoting angles θ as shown in Figure 6.2(b), the values for angles *a, b, c* and *d* are obtained as follows:

$$a = 90° - \theta$$

In terms of *a*, angles $b = a/2$ (as they subtend the same arc lengths as angles *a* but extend to the perimeters of the circles); angle $c = 90° - b$; and at intersection point 1, angle $d = c - b$. Thus, in terms of θ:

1. The angle *b* between the direction of the discontinuity and the plane of the major principal stress in the lower stressed zone (zone B) is given by

$$b = \frac{90° - \theta}{2} \tag{6.5}$$

2. The angle *c* between the direction of the discontinuity and the plane of the major principal stress in the higher stressed zone (zone C) is given by

$$c = \frac{90° + \theta}{2} \tag{6.6}$$

3. The angle *d* between the planes on which σ_{1B} and σ_{1C} act is given by

$$d = \theta \tag{6.7}$$

As the radii of the two circles are c_u it follows from the equilateral triangle $s_B M s_C$ in Figure 6.2(b) that the change in the stress parameter *s* across the discontinuity is given by

$$\Delta s = s_C - s_B = 2c_u \sin \theta \tag{6.8}$$

6.4 Earth pressure on a rough retaining wall

In Section 6.2 the pressure on a smooth vertical retaining wall was found without the need to insert a stress discontinuity in the soil behind the wall, because the directions of principal stresses were vertical and horizontal in the soil mass and against the wall. If it is assumed that shear stress will develop between the soil and the wall, it is necessary to insert a stress discontinuity to cope with the change in principal stress direction.

In considering the active case the difficulty again arises, as discussed in Section 6.2, that the calculations give tensile stresses acting at the top of the wall. A solution can again be obtained by assuming the cracked soil to a depth of $2c_u/\gamma$ simply acts as a surcharge on the soil below this depth, but the problem of making an allowance for water pressure in the cracks leads inevitably to a hybrid and unsatisfactory solution. The passive case is represented in Figure 6.3(a), where it is assumed that

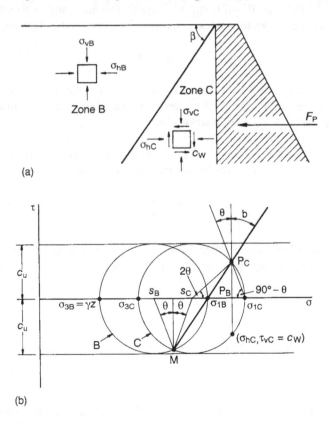

Figure 6.3 Rough retaining wall, passive case; (a) stress discontinuity dividing zones B and C; (b) total stress circles.

the shear stress generated between the soil and the wall is given by c_W where c_W $\leqslant c_u$. It is possible to plot a point (σ_{hC}, τ_{vC}) a distance c_W below the abscissa on a Mohr stress diagram and complete a unique circle through this point tangential to the $\pm c_u$ envelopes (circle C in Figure 6.3(b)). The pole point P_C is found by projecting vertically from the stress point, and the direction of the plane on which σ_{1C} acts is given by a line through P_C and σ_{1C}. As the direction of the plane on which σ_{hB} $(= \sigma_{1B})$ acts is vertical, the angle θ is the angle of the line $P_C\sigma_{1C}$ to the vertical (equation 6.7). The angle θ is given by

$$\sin 2\theta = \frac{c_W}{c_u} \tag{6.9}$$

as seen in Figure 6.3(b).

A line drawn at an angle θ to the vertical from the centre of stress circle C (cf. Figure 6.2) gives the intersection point of the stress circles for zones B and C, and thus allows the stress circle B to be drawn. As the minor principal stress for circle B is equal to the vertical stress γz in zone B, the diagram can be completed quantitatively for specific values of γz.

The direction of the discontinuity for the specific value of c_W assumed is given by a straight line through MP_BP_C in Figure 6.3(b), and the angle b between the discontinuity and the plane of the major principal stress in the lower stressed zone (zone B) is as given by equation 6.5.

Taking a simple example, assuming $c_W = 0.5c_u$, gives $\theta = 15°$ from equation 6.9 and $b = 37.5°$. As the plane of the major principal stress is vertical in zone B the discontinuity angle β in Figure 6.3(a) is thus 52.5°.

The horizontal stress against the wall at depth z is equal to σ_{hC} which has the magnitude (Figure 6.3(b)) of

$$\sigma_{hC} = \gamma z + c_u + (S_C - S_B) + c_u \cos 2\theta \tag{6.10}$$

where $(s_C - s_B)$ is given by equation 6.8, thus

$$\sigma_{hC} = \gamma z + c_u (1 + 2 \sin\theta + \cos 2\theta) \tag{6.11}$$

The total passive horizontal force F_p on a wall of height H is therefore

$$F_p = \int_0^H \sigma_{hC} \, dz$$
$$= \frac{\gamma H^2}{2} + c_u H(1 + 2\sin\theta + \cos 2\theta) \tag{6.12}$$

EXAMPLE 6.1 ACTIVE FORCE ON RETAINING WALL DUE TO SURCHARGE

Although total stress calculations are unsatisfactory for calculating active pressures generated by the soil itself behind retaining walls, this approach can be useful in calculating wall pressures arising from surcharge loading on the soil surface as shown in Figure 6.4(a). The retaining wall in Figure 6.4(a) supports a 3 m depth of soft clay with an undrained shear strength $c_u = 20$ kPa. If a surcharge q_s of 80 kPa is placed on the surface of the clay, find the horizontal force F_{hW} per metre length of wall acting on the face AB if: (a) the adhesion c_W between the soil and wall is zero; (b) $c_W = 0.9c_u$. Assume the soil to be weightless.

Solution

(*a*) $c_W = 0$

As $c_W = 0$, the horizontal stress σ_{hW} on the wall is a principal stress and thus, as shown in Figure 6.4(b):

$$\sigma_{hW} = q_s - 2c_u = 40 \text{ kPa}$$

$$\therefore F_{hW} = 3 \times 40 = 120 \text{ kN/m}$$

(*b*) $c_W = 0.9c_u$

The steps in solving this case are as follows.

1. In order to find the direction of the major principal stress σ_{1B} acting in zone B, and hence acting on the wall, reference is made to Figure 6.4(d). Point N represents the stress conditions on the vertical face of a soil element adjacent to the wall as shown in Figure 6.4(c). It is positive, as the soil is tending to move downwards relative to the wall and consequently c_W acts in an anticlockwise direction on the vertical faces of the element. As it is an active stress condition σ_{hW} will be less than s_B.
2. Pole point for planes P_B in Figure 6.4(d) is found by projecting vertically from point N to intersect the stress circle. The line $P_B\sigma_{1B}$ then gives the directions of planes in zone B on which σ_{1B} acts. From the geometry of Figure 6.4(d) these planes are at 32° to the horizontal, and thus the major principal stress changes in direction by 32° from zone C to zone B.
3. In Figure 6.4(e) the stress circle for zone C is known, as it has radius equal to c_u and passes through $\sigma_{1C} = q_s = 80$ kPa. By constructing the equilateral triangle $s_C M s_B$, with $\theta = 32°$, the centre of circle B is established. The value of σ_{hW} can now be found from the geometry of Figures 6.4(d) and 6.4(e):

$$\sigma_{hW} = s_B - c_u \cos 2\theta$$
$$s_B = s_C - 2c_u \sin \theta$$
$$= q_S - c_u(1 + 2\sin \theta)$$
$$\therefore \sigma_{hW} = q_S - c_u(1 + 2\sin \theta + \cos 2\theta)$$

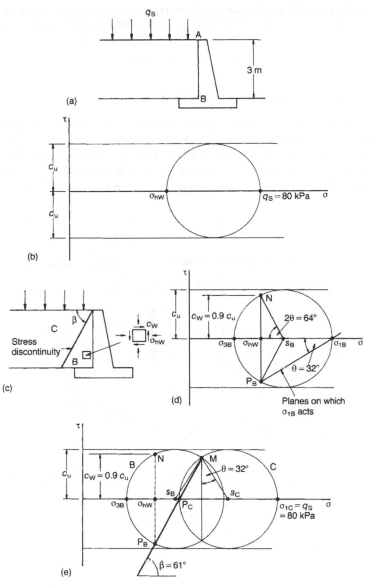

Figure 6.4 Example 6.1.

Substituting $q_S = 80$ kPa, $c_u = 20$ kPa gives

$$\sigma_{hW} = 30\,\text{kPa}$$
$$\therefore F_{hW} = 90\,\text{kN/m}$$

The angle β of the discontinuity to the horizontal can be found by a line through the common stress point M for zones B and C and the pole points P_B, P_C. From equation 6.5 and the geometry of Figure 6.4(e) this gives $\beta = 61°$.

6.5 Foundation with smooth base

A simple solution to the bearing capacity of a surface footing of infinite length with a smooth base can be obtained by assuming a weightless soil, with vertical discontinuities below each edge of the footing as shown in Figure 6.5(a). The resulting stress circles are those shown in Figure 6.5(b) and, as $\sigma'_{3B} = p_S$, the surface pressure, the lower bound failure loading q_F is given by

$$q_F = \sigma_{1A} = 4c_u + p_S \qquad (6.13)$$

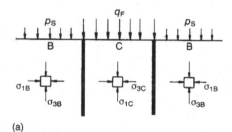

(a)

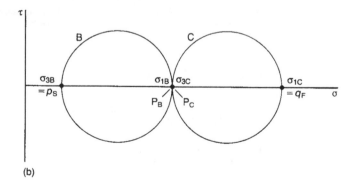

(b)

Figure 6.5 Foundation with smooth base: (a) vertical discontinuities; (b) total stress circles.

This compares with the 'correct' solution

$$q_F = 5.14c_u + p_S \tag{6.14}$$

A better solution can be obtained by assuming two or more discontinuities radiating from each corner, as shown in Figures 6.6(a)–(c). The number of discontinuities required for a solution simply equals the number of stress changes decided upon, which also equals the number of changes in direction of the major (and minor) principal stress. The larger the number of discontinuities, the better the solution. In the solution above with a single discontinuity, the direction of the major principal stress changes by 90° across one discontinuity. If two discontinuities are assumed the changes in direction across them must total 90°, and if the changes across each are assumed equal (which is not a necessary assumption, but leads to symmetry in the disposition of the discontinuities) the change in direction across each must be 45°.

Two discontinuities from each corner divide the foundation soil into three zones, B, C and D. The steps in constructing the three stress circles for these zones, shown in Figure 6.6(d), are as follows.

1. As p_S is known, circle B can be drawn with centre

$$s_B = p_S + c_u$$

    and radius c_u, where c_u is the undrained shear strength.
2. Pole point for planes p_B is found by projecting a horizontal line from p_S to intersect circle B.
3. As the major principal stress rotates through 90° from zone B to zone C, and two discontinuities are assumed, the rotation across each discontinuity is 45°, and the stress point N, common to zones B and D, is found by drawing line $s_B N$ such that $\theta = 45°$.
4. The centre s_D of the stress circle D for zone D is found by completing the equilateral triangle $s_B N s_D$. The stress circle D can then be drawn with radius c_u passing through point N.
5. The orientation of the plane on which stress N acts in zone B is found by a line through N and P_B. As this stress is continuous across the discontinuity separating zones B and D, extension of line NP_B to intersect circle D gives pole point P_D. The slope β of line $NP_B P_D$ can readily be found from the geometry of the figure to be $\beta = 67.5°$. This is the slope of the discontinuity separating zones B and D.
6. Projection of the line $s_D M$ at $\theta = 45°$ locates the common stress point M for zones D and C, and completion of the equilateral triangle $s_D M s_C$ locates s_C, the centre of stress circle C, which can now be drawn, passing through point M. Pole point P_C is found by projecting horizontally from stress point σ_{1B} to

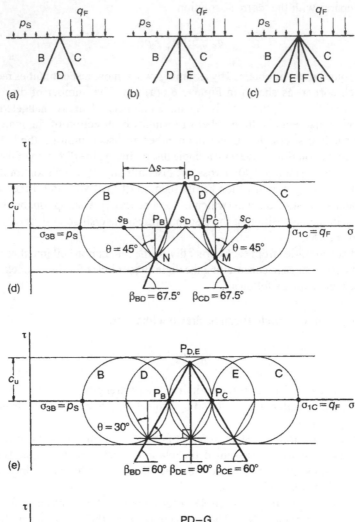

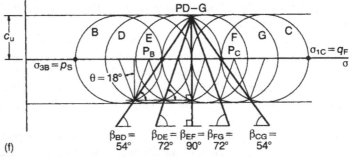

Figure 6.6 Foundation with smooth base, solutions assuming: (a,d) two discontinuities; (b,e) three discontinuities; (c,f) five discontinuities.

intersect the stress circle C. A line through $P_D P_C M$ gives the slope $\beta = 67.5°$ of the discontinuity separating stress zones D and C.

7. It can be seen in Figure 6.6(d) that $q_F = \sigma_{1B}$ is given in terms of p_s and c_u as follows:

$$q_F = p_S + 2c_u + 2\Delta s$$

but

$$\Delta s = s_C - s_D = s_D - s_B = 2c_u \sin\theta$$

$$\therefore q_F = p_S + 2c_u\left(1 + 2\sin\frac{\pi}{4}\right) \tag{6.15}$$

The stress diagrams for three and five discontinuities shown in Figures 6.6(e) and 6.6(f) can be drawn adopting exactly the same principles as for two discontinuities as described above, but advancing the circles each time by $90°/n$, or $\pi/2n$ radians, where n is the number of assumed discontinuities. Thus $\theta = 30°$ for three discontinuities in Figure 6.6(e) and $\theta = 18°$ for five discontinuities in Figure 6.6(f). It will be seen from Figures 6.6(d)–(f) that the general expression relating q_F to p_S and c_u is

$$q_F = p_S + 2c_u\left(1 + n\sin\frac{\pi}{2n}\right) \tag{6.16}$$

or

$$q_F = p_S + N_c c_u \tag{6.17}$$

where

$$N_c = 2\left(1 + n\sin\frac{\pi}{2n}\right) \tag{6.18}$$

Values of N_c for $n = 1, 2, 3$ and 5 are given in Table 6.1, together with the value for $n = \infty$ from Section 8.2.1.

It can be seen that two discontinuities give a marked improvement in the N_c value compared to one discontinuity, while three discontinuities give a value only 3% less than the true value of 5.14, which is well within the accuracy to which c_u can be determined.

A fully valid lower bound solution requires that the conditions of stress equilibrium and non-violation of the yield or failure criterion are met at all points within the deforming mass. If two discontinuities from both corners of a surface loading are extended to unlimited depths, as shown in Figure 6.7(a), then the two discontinuities which cross form a further zone E, within which the stress

Table 6.1 Values of N_c

No. of discontinuities	N_c
1	4.0
2	4.83
3	5.00
5	5.09
∞	5.14 (Section 8.2.1)

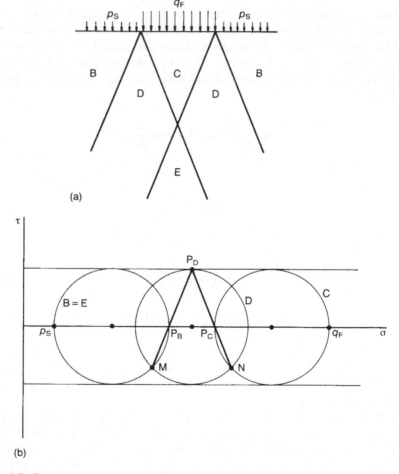

(a)

(b)

Figure 6.7 Check on stresses in zone E.

equilibrium and failure criterion must not be violated. The stress circle for this zone can be easily found by projecting a line from P_D in Figure 6.7(b) parallel to the discontinuity separating zones D and E to give the interaction of the E circle with the D circle. It is seen that this gives a stress circle for zone E coincident with the circle for zone B.

EXAMPLE 6.2 BEARING CAPACITIES UNDER STRIP LOADINGS

The strip loadings shown in Figures 6.8(a) and (b) are 1 m wide and rest on clay, assumed to be weightless, with an undrained shear strength of 100 kPa. Calculate the bearing capacities q_F for the geometries shown, assuming in each case two discontinuities as appropriate radiating from the edges of the applied loadings.

Solution

Case (a)

As the surface of the 20° slope is stress free and the soil is weightless, the principal stresses in zone B are parallel and normal to the direction of the slope. The minor principal stress σ_{3B} is zero, and consequently the stress circle B in Figure 6.8(c) can be drawn. The pole point P_B can then be located. As the major principal stress rotates through 70° from zone B to zone C, and two discontinuities are being assumed, the common stress point N for zones B and D is found by a line drawn from s_B with $\theta = 35°$. Completion of the equilateral identical triangles $s_B N s_D$ and $s_D M s_C$ allows stress circles D and C to be drawn. Pole point P_C is found by projecting horizontally from stress point σ_{1C} to intersect the circle. Lines through N and P_B and through M and P_C intersect at pole point P_D on stress circle D. From the geometry of Figure 6.8(c):

$$q_F = \sigma_{1C} = 2c_u + 4c_u \sin 35°$$

$$= 429 \text{ kPa}$$

It can also be shown that the two discontinuities have slopes of 81.5° and 62.5° to the horizontal.

Case (b)

The procedure for case (b) is identical to that for case (a), but the major principal stress now rotates through 110°, giving $\theta = 55°$. This gives the stress diagram shown in Figure 6.8(d), from which

$$q_F = \sigma_{1C} = 2c_u \sin 55°$$
$$= 528 \text{ kPa}$$

The two discontinuities have slopes of 52.5° and 72.5° to the horizontal as shown.

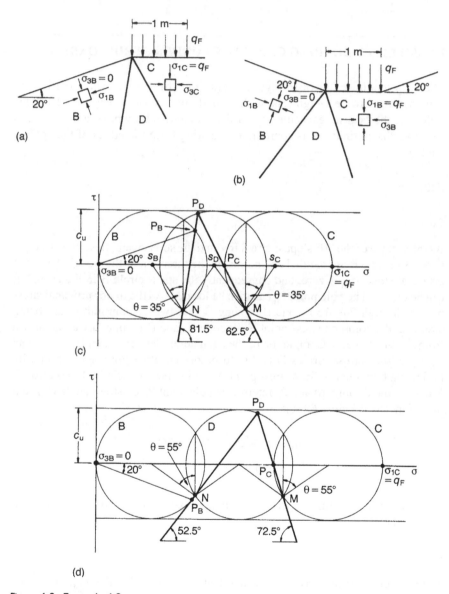

(a)

(b)

(c)

(d)

Figure 6.8 Example 6.2

6.6 Undrained flow between rough parallel platens

The flow of a plastic material squeezed between parallel platens, as shown in Figure 6.9, has practical significance in relation to a thin layer of clay separating a foundation from an underlying rock surface. It also has significance in relation to temporary mine support systems. Where a plate, loaded hydraulically or otherwise, is used to support a portion of an underground excavation it is possible for a piece of loose rock to be caught between the plate and the rock surface. If this piece of rock flows as a plastic material, extremely high localized pressures can be generated leading to buckling of the support plate. This development of an intense localized pressure is also used in the manufacture of artificial diamonds.

A layer of saturated clay compressed between parallel rough platens is shown in Figure 6.10(a). A restraining pressure σ_s acts on the end faces and plane strain conditions are assumed to obtain normal to the paper. A lower bound solution to the distribution of pressure between the platens and the clay can be found by dividing the deforming soil into regions a, b, c, d, etc., separated by planar discontinuities having angles alternately at 60° and 30° to the horizontal. The asymmetry of each of these 'kite shaped' zones about a vertical axis reflects the flow of the deforming soil to the right. The direct stresses between the platens and the clay are denoted σ_{vb}, σ_{ve}, σ_{vh}, etc. and the corresponding shear stresses τ_{hb}, τ_{he}, τ_{hh}, etc.

In zone a, the minor principal stress is horizontal equal to σ_s, and the major principal stress is vertical. This gives the stress circle a, shown in Figure 6.10(b), with the pole point for planes P_a at the stress point σ_s. The stresses along discontinuities separating zones a and b, and zones a and c, are found by projecting lines from P_a parallel to the discontinuities to meet circle a, which then allows the single stress circle for zones b and c to be drawn. Pole points P_b, P_c coincide with the stress intersection points. The stresses in successive regions a, b/c, d, e/f, etc.

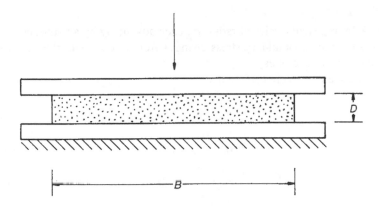

Figure 6.9 Plastic material squeezed between parallel platens.

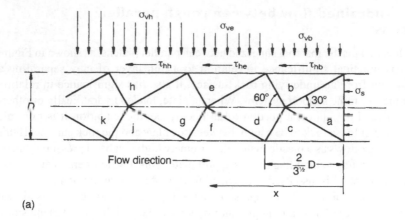

(a)

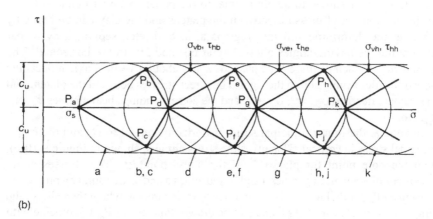

(b)

Figure 6.10 Undrained solution for a saturated clay squeezed between parallel platens: (a) geometry of discontinuities; (b) total stress circles.

are given by repeated circles of radius c_u, each advancing by an amount c_u along the horizontal axis. Boundary stress changes from one zone to the next can be obtained from these circles.

It can be seen in Figure 6.10(b) that

$$\sigma_{vb} = \sigma_s + 2.5c_u \qquad (6.19a)$$

$$\sigma_{ve} = \sigma_s + 4.5c_u \qquad (6.19b)$$

$$\sigma_{vh} = \sigma_s + 6.5c_u \qquad (6.19c)$$

That is, σ_v increases by $2c_u$ from one boundary region to the next. Also

$$\tau_{hb} = \tau_{he} = \tau_{hh} = 1.5c_u \tan 30^{\circ}$$

i.e.

$$\tau_h = 0.866c_u \qquad\qquad (6.20)$$

Thus, the shear stress is constant along the boundary, with a value for this solution slightly less than the undrained shear strength of the soil.

In Figure 6.11 the normal stress σ_v between the platens and the soil is plotted in the dimensionless form $(\sigma_v - \sigma_s)/c_u$ vs x/D, where x is the distance from the free boundary, as shown in Figure 6.10(a), and D is the thickness of the deforming soil.

Assuming a deforming strip of width B, as shown in Figure 6.9 (with plane strain conditions normal to the paper), the maximum stepped pressure in the centre of the strip can be seen from Figure 6.11 to have the values shown in Table 6.2. The average pressures are also shown.

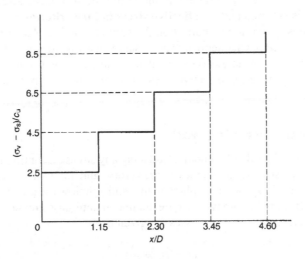

Figure 6.11 Distribution of normal stress on platens.

Table 6.2 Maximum and average pressures in the centre of a deforming strip of width B

$\dfrac{B}{D}$	$\dfrac{\sigma_{vmax} - \sigma_s}{c_u}$	$\dfrac{\sigma_{v\,av} - \sigma_s}{c_u}$
6.9	6.5	4.5
9.2	8.5	5.5
18.4	16.5	9.5

Chapter 7

The use of stress discontinuities in drained plasticity calculations

7.1 Lower bound drained solutions

If a soil at failure is assumed to be deforming plastically with full dissipation of excess pore pressure permitted, simple solutions to stability problems can often be obtained by ensuring that, at any point within the soil mass, **effective stress equilibrium** is satisfied and the **effective stress failure criterion** is not violated. These solutions give a **lower bound** on the applied loading. As with undrained solutions, discussed in Chapter 6, it is necessary to assume for most problems one or more changes in the stress field to satisfy boundary conditions. This can be accomplished by inserting effective stress discontinuities within the deforming mass.

7.2 Smooth retaining wall

This problem can be solved without introducing any stress discontinuities because the principal stresses throughout the soil are vertical and horizontal. If the bulk unit weight of the soil is γ_a above the phreatic line and γ_b below the phreatic line, then for depths z_a (Figure 7.1) above the phreatic line, if capillary stresses are ignored, the total and effective vertical stresses are given by

$$\sigma_v = \sigma_v' = \gamma_a z_a \tag{7.1}$$

and for depths z_b below the phreatic line,

$$\sigma_v = \gamma_a z_w + \gamma_b (z_b - z_w) \tag{7.2a}$$

$$\sigma_v' = \gamma_a z_w + (\gamma_b - \gamma_w)(z_b - z_w) \tag{7.2b}$$

The total lateral stress, and hence the total pressure on the wall, is given by:

Above the phreatic line: $\qquad \sigma_{hW} = \sigma_{hW}' \tag{7.3}$

Below the phreatic line: $\quad \sigma_{hW} = \sigma'_{hW} + \gamma_W(z_b - z_W)$ $\hspace{2cm}$ (7.4)

where σ'_{hW} can be shown from Figure 7.1b, c to be given by the expressions:

Active case: $\hspace{1cm}$ $\sigma'_{hW} = \sigma'_v\left(\dfrac{1-\sin\phi'}{1+\sin\phi'}\right) - c'\left(\dfrac{2\cos\phi'}{1+\sin\phi'}\right)$ $\hspace{1.5cm}$ (7.5)

Passive case: $\hspace{1cm}$ $\sigma'_{hW} = \sigma'_v\left(\dfrac{1+\sin\phi'}{1-\sin\phi'}\right) + c'\left(\dfrac{2\cos\phi'}{1+\sin\phi'}\right)$ $\hspace{1.5cm}$ (7.6)

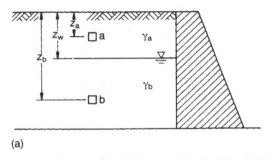

(a)

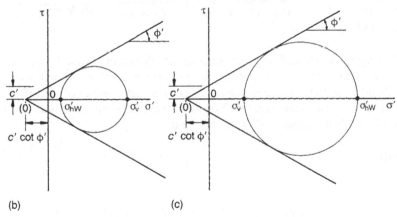

(b) $\hspace{5cm}$ (c)

Figure 7.1 Smooth retaining wall: (a) soil and ground water conditions; (b) effective stress circle for active case; (c) effective stress circle for passive case.

In order to derive equations 7.5 and 7.6 from Figures 7.1(b) and 7.1(c) it is convenient to establish a false origin (0) by projecting the failure envelopes back to the abscissa, where they intersect at a distance of $-c'\cot\phi'$ from the true origin. This device was used by Sokolovski (1965).

At shallow depths below ground level, equation 7.5 gives a negative (tensile) pressure. The depth to which this occurs can be found by putting $\sigma'_{hw} = 0$ in equation 7.5 and substituting the appropriate value of σ'_v from equation 7.1 or equation 7.2b. Thus, the calculated depth of the tension crack depends upon the depth of the water table.

In many soils it is prudent to assume $c' = 0$, in which case the above problem does not arise. If c' is taken to have a finite value, and the water table is below ground level, it will usually be advisable to assume hydraulic pressure acting within the depth of the calculated tensile crack, unless specific measures are taken in the field to relieve such pressures.

If $c' = 0$, equations 7.5 and 7.6 reduce to

Active case:
$$\sigma'_{hw} = \sigma'_v \left(\frac{1 - \sin \phi'}{1 + \sin \phi'} \right) \tag{7.7}$$

Passive case:
$$\sigma'_{hw} = \sigma'_v \left(\frac{1 + \sin \phi'}{1 - \sin \phi'} \right) \tag{7.8}$$

EXAMPLE 7.1 ACTIVE AND PASSIVE FORCES ON SMOOTH RETAINING WALL

A 3 m high smooth retaining wall such as that shown in Figure 7.1(a) supports sand fill with a dry unit weight of 16.5 kN/m³ and a saturated unit weight of 20 kN/m³. Find the active and passive forces per metre length of wall if: (a) the sand is dry; (b) the water table is at the sand surface. Assume the unit weight of water is 10 kN/m³. Take $\phi' = 30°$ for the sand.

Solution

(a) Dry sand

$$\text{Equation 7.1: } \sigma'_v = \sigma_v = \gamma z = 16.5z \text{ kPa}$$

Active pressure – putting $\phi' = 30°$ into equation 7.7 gives

$$\sigma'_{hw} = \tfrac{1}{3} \gamma z = 5.5z \, \text{kPa}$$

$$F_{hw} = \tfrac{1}{3} \gamma \frac{H^2}{2}$$

where H is the height of the wall,

$$\therefore F_{hW} = 24.8\,\text{kN/m}$$

Passive pressure – putting $\phi' = 30°$ into equation 7.8 gives

$$\sigma'_{hw} = 3\gamma z = 49.5z\ \text{kPa}$$

$$F_{hW} = 3\gamma\frac{H^2}{2}$$

$$= 233\,\text{kN/m}$$

(b) Water table at the soil surface

$$\text{Equation 7.2b: } \sigma'_v = \gamma'z = (\gamma_b - \gamma_w)z = 10z\ \text{kPa}$$

Active pressure – putting $\phi' = 30°$ into equation 7.7 gives

$$\sigma'_{hw} = \tfrac{1}{3}\gamma'z = 3.33z\ \text{kPa}$$

$$\therefore F_{hW} = 3.33 \times \frac{3^2}{2} = 15\,\text{kN/m}$$

Add water pressure:

$$F_{hW} = 15 + 10 \times \frac{3^2}{2} = 60\,\text{kN/m}$$

Passive pressure – putting $\phi' = 30°$ into equation 7.8 gives

$$\sigma'_{hw} = 3\gamma z = 30z\ \text{kPa}$$

$$\therefore F'_{hW} = 30 \times \frac{3^2}{2} = 135\,\text{kN/m}$$

Add water pressure:

$$F_{hW} = 135 + 45 = 180\,\text{kN/m}$$

7.3 Effective stress discontinuity

The effective stress conditions on either side of the discontinuity in Figure 7.2(a) are represented by the stress circles B and C in Figure 7.2(b), which satisfies the following stress conditions:

$$\sigma_{nC} = \sigma'_{nB} \tag{7.9a}$$

$$\tau_{ntC} = \tau_{ntB} \tag{7.9b}$$

$$\sigma'_{tC} \neq \sigma'_{tB} \tag{7.9c}$$

The lower intersection point M of the two circles represents the direct and shear stresses across the discontinuity. A line through the false origin (O) and M has the angle δ' to the horizontal, and it is convenient also to introduce the angle Δ as shown. Various Sections from Figure 7.2(b) have been isolated and reproduced in Figures 7.2(c) to 7.2(g) to present the geometry more clearly.

From Figure 7.2(c):

$$s'_B + c' \cot \phi' = \frac{R_B}{\sin \phi'} \tag{7.10}$$

From Figure 7.2(d):

$$s'_B + c' \cot \phi' = \frac{R_B \sin(180° - \Delta)}{\sin \delta'} \tag{7.11}$$

From equations 7.10 and 7.11:

$$\sin \Delta = \frac{\sin \delta'}{\sin \phi'} \tag{7.12}$$

As ϕ' and δ' are known, Δ can be found from equation 7.12.

A line drawn through the intersection point M parallel to the discontinuity establishes the pole points for the planes, P_B and P_C, in Figure 7.2(b). The directions of the planes on which the major principal stresses act in the two zones are given by straight lines drawn through $P_B \sigma'_{1B}$ for zone B and $P_C \sigma'_{1C}$ for zone C. It follows from the geometry of Figure 7.2(b) (also refer to Figures 7.2(e) and (f)) that:

1. The angle b between the direction of the discontinuity and the plane of action of the major principal stress in the lower stressed zone (zone B) is

$$b = \tfrac{1}{2}(\Delta + \delta') \tag{7.13}$$

2. The angle c between the direction of the discontinuity and the plane of the major principal stress in the higher stressed zone (zone C) is

$$c = \tfrac{1}{2}(180° - \Delta + \delta') \qquad (7.14)$$

3. The angle θ between the planes of σ'_{1B} and σ'_{1C} is given by

$$\theta = c - b = 90° - \Delta \qquad (7.15)$$

Thus, the major principal stresses change direction by the angle $(90° - \Delta)$ across a discontinuity. It should be noted that all the expressions above also hold for the case of $c' = 0$.

The change in stress state s'_B to s'_C across the discontinuity can be found by referring to Figure 7.2(g):

$$\frac{s'_B + c' \cot \phi'}{\sin \Delta} = \frac{(O)M}{\sin(180° - \Delta - \delta')}$$

$$\frac{s'_C + c' \cot \phi'}{\sin(180° - \Delta)} = \frac{(O)M}{\sin(\Delta - \delta')}$$

Eliminating $\sin \Delta = \sin(180° - \Delta)$ and $(O)M$ gives

$$\frac{s'_C + c' \cot \phi'}{s'_B + c' \cot \phi'} = \frac{\sin(\Delta + \delta')}{\sin(\Delta - \delta')} \qquad (7.16)$$

If $c' = 0$:

$$\frac{s'_C}{s'_B} = \frac{\sin(\Delta + \delta')}{\sin(\Delta - \delta')} \qquad (7.17)$$

Referring to Figure 7.2(b), the ratio of σ'_{1C} to σ'_{1B} is given by

$$\frac{\sigma'_{1C}}{\sigma'_{1B}} = \frac{s'_C(1 + \sin \phi')}{s'_B(1 - \sin \phi')}$$

Thus

$$\frac{\sigma'_{1C}}{\sigma'_{1B}} = \left(\frac{\sin(\Lambda + \delta')}{\sin(\Delta - \delta')} \right) \left(\frac{1 + \sin \phi'}{1 - \sin \phi'} \right) \qquad (7.18)$$

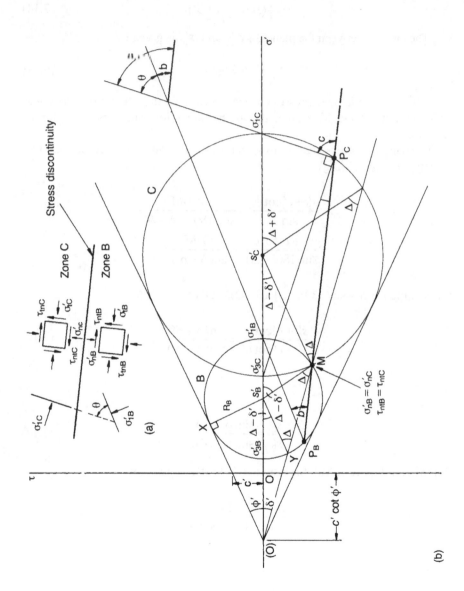

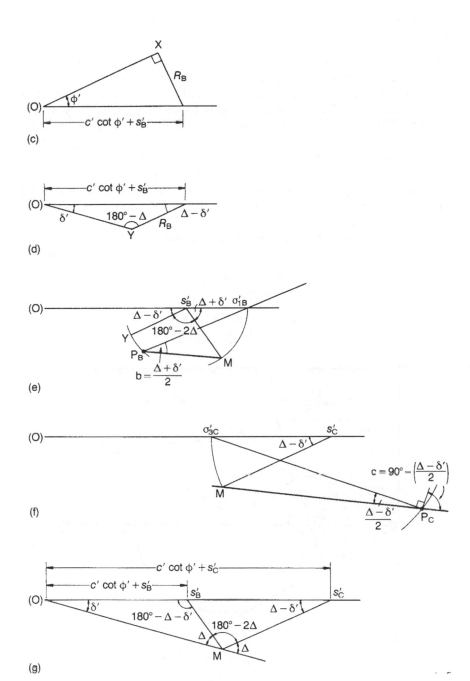

Figure 7.2 (a) Stress discontinuity; (b) corresponding effective stress circles; (c–g) isolated Sections of stress diagram.

7.4 Active earth pressure on a rough retaining wall

The effect of a rough wall is to give a calculated active normal pressure on the wall σ'_{hW} less than that for a smooth wall, but the reduction is not large even for assumed values of wall friction ω' of the same order of magnitude as the friction angle ϕ' for the soil. The solution for $\omega' = \phi'$ is shown in Figure 7.3(b), assuming $\phi' = 30°$, $c' = 0$.

In Figure 7.3(b) point X represents the stresses at the wall face and circle B through point X the stresses in soil zone B. A vertical line from X establishes the pole point for planes P_B. The line $P_B\sigma'_{1B}$ is the direction of the plane of action of the major principal stress in zone B and, from equation 7.15, this makes the angle $(90° - \Delta)$ with the direction of the plane of major principal stress in zone C, which is horizontal.

Thus, from the geometry of Figure 7.3 (and equation 7.15):

$$\theta = 90° - \Delta = 30°$$

$$\therefore \quad \Delta = 60°$$

Equation 7.12: $\sin \delta' = \sin \Delta \cdot \sin \phi'$

$$\therefore \quad \delta' = 25.66°$$

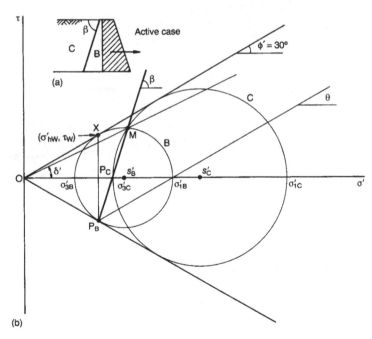

Figure 7.3 Rough retaining wall, active case: (a) stress discontinuity; (b) effective stress circles.

A line with slope angle δ' can now be drawn through the origin O to intersect circle B at point M. This is the point of intersection of the two stress circles B and C, so circle C can now be drawn.

From equation 7.14 the angle of the discontinuity is given by

$$\beta = \tfrac{1}{2}(180° - \Delta + \delta')$$
$$\therefore \quad \beta = 72.8°$$

Equation 7.17: $\quad \dfrac{s'_C}{s'_B} = \dfrac{\sin(\Delta + \delta')}{\sin(\Delta - \delta')}$

$$\therefore \quad \frac{s'_C}{s'_B} = 1.768$$

Thus, from Figure 7.3(b):

$$\sigma'_{hW} = s'_B \cos^2 \phi' = 0.75 s'_B$$
$$\sigma'_{1C} = \gamma z = s'_C(1 + \sin \phi') = 1.5 s'_C$$
$$\therefore \quad \sigma'_{hW} = 0.28\gamma z$$
$$\tau_W = \sigma'_{hW} \tan \phi'$$
$$\therefore \quad \tau_W = 0.16\gamma z$$

The value of $\sigma'_{hW} = 0.28yz$ compares with $\sigma'_{hW} = 0.33yz$ for the smooth wall in Example 7.1.

7.5 Passive earth pressure on a rough retaining wall

The effect of a rough wall is to give a calculated passive normal pressure on the wall σ'_{hW} greater than that for a smooth wall. Figure 7.4(a) shows such a wall with an assumed friction angle ω' between the wall and the soil.

In Figure 7.4(b) the stress point σ'_{hW}, τ_W lies along a line at an angle ω' to the horizontal as shown. τ_W is negative because the soil is tending to move upwards and thus τ_W is clockwise adjacent to the wall. With the point σ'_{hW}, τ_W located, a stress circle C, representing the stresses in zone C, can be drawn passing through this point and touching the failure envelopes. The pole point P_C can now be located by projecting vertically from point σ'_{hW}, τ_W to intersect circle C at point P_C.

By analogy with equation 7.12, the angle Ω in Figure 7.4(b) is given by

$$\sin\Omega = \frac{\sin\omega'}{\sin\phi'} \tag{7.19}$$

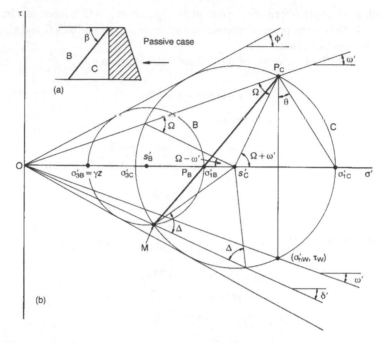

Figure 7.4 Rough retaining wall, passive case: (a) stress discontinuity; (b) effective stress circles.

The angle subtended at the centre of circle C by the arc $\sigma'_{3C}P_C$ is

$$(180° - 2\Omega) + (\Omega - \omega') = 180° - \Omega - \omega'$$

Thus, the angle of the line $P_C\sigma'_{1C}$ to the horizontal, which is also subtended by the arc $\sigma'_{3C}P_C$ is ½ $(180° - \Omega - \omega')$. In zone B the major principal stress acts on a vertical plane, so the angle between the major principal stress directions in zones B and C is given by

$$\theta = 90° - \tfrac{1}{2}(180° - \Omega - \omega')$$
$$= \tfrac{1}{2}(\Omega + \omega') \tag{7.20}$$

It is convenient now to proceed with assumed values of ϕ' and ω'. Assuming $\phi' = 30°$ and $\omega' = 20°$,

Equation 7.19: $\Omega = 43.16°$

Equation 7.20: $\theta = 31.58°$

Equation 7.15: $\Delta = 58.42°$

Equation 7.12: $\delta = 25.21°$

A line drawn at angle δ' to the horizontal as shown in Figure 7.4(b) meets stress circle C at M, the point of intersection of stress circles B and C, thus allowing stress circle B to be drawn. As $\sigma'_{3B} = \gamma z$, the stresses against the wall can now be evaluated.

$$\text{Equation 7.17: } \frac{s'_C}{s'_B} = \frac{\sin(58.42° + 25.21°)}{\sin(58.42° - 25.21°)}$$

$$= 1.814$$

From Figure 7.4(b):

$$\sigma'_{hW} = s'_C[1 + \tfrac{1}{2}\cos(\Omega + \omega')]$$

$$= 1.255 s'_C$$

$$s'_B = \frac{\sigma'_{3B}}{1 - \sin\phi'} = 2\gamma z$$

$$\therefore \quad \sigma'_{hW} = 4.44\gamma z$$

$$\tau_W = 4.44\gamma z \tan\omega'$$

$$= 1.6\gamma z$$

The value of $\sigma'_{hW} = 4.44$ compares with $\sigma'_{hW} = 3.0$ for the smooth wall in Example 7.1.

Referring to Figure 7.4(b), the slope of the discontinuity is given by the line passing through $P_C P_B M$. This has a slope, from equation 7.13, of $\tfrac{1}{2}(\Delta + \delta')$ to the vertical, and thus

$$\beta = 48.2°$$

EXAMPLE 7.2 PASSIVE FORCE ON ROUGH RETAINING WALL

The 3 m high rough retaining wall in Figure 7.5(a) supports a slightly clayey sand fill with a saturated unit weight of 20 kN/m³. The effective stress soil strength properties are $c' = 10$ kPa, $\phi' = 30°$, and the wall friction $\omega' = \phi'/2$. Find the passive force per unit length of wall if the water table is at the soil surface and $\gamma_w = 10$ kN/m³. Assume a single stress discontinuity and find the angle β to the horizontal of the discontinuity.

Solution

As $z_w = 0$ and $z_b = z$, equation 7.2(b) can be written, for zone B:

$$\sigma'_{vB} = \gamma'z$$

where γ' is the buoyant unit weight of the soil.

$$\therefore \ \sigma'_{vB} = \sigma'_{3B} = 10z \text{ kPa} \ (z \text{ in metres})$$

Putting $\omega' = 15°$, $\phi' = 30°$ into equation 7.19 gives

$$\sin\Omega = \frac{\sin 15°}{\sin 30°}$$

$$\therefore \ \Omega = 31.2°$$

The angle θ of rotation of the major principal stress from zone B to zone C is given by equation 7.20, thus

$$\theta = \tfrac{1}{2}(31.2° + 15°)$$

$$= 23.1°$$

It follows from equations 7.15 and 7.12 that

$$\theta = 90° - \Delta = 66.9°$$

$$\therefore \ \delta' = 27.4°$$

$$\text{Equation 7.16:} \ \frac{s'_C + 10\cot 30°}{s'_B + 10\cot 30°} = \frac{\sin(66.9° + 27.4°)}{\sin(66.9° - 27.4°)}$$

From which

$$s'_C = 1.56 s'_B + 9.7 \text{ kPa}$$

But, from the geometry of Figure 7.5(b), for $\phi' = 30°$:

$$s'_B = 2\sigma'_{3B} + c'\cot\phi'$$

$$= 20z + 17.3 \text{ kPa}$$

$$\therefore \ s'_C = 31.2z + 36.7 \text{ kPa}$$

From Figure 7.5(b)

$$\sigma'_{hw} = s'_C + R_C \cos(\Omega + \omega')$$

where

$$R_C = \text{radius of circle C}$$
$$= 0.5 s_C' + 8.66 \text{ kPa}$$
$$\therefore \sigma_{hW}' = 42z + 55.4 \text{ kPa}$$
$$\text{and } F_{hW}' = 42 \times \frac{3^2}{2} + 55.4 \times 3 \text{ kN/m}$$
$$= 355 \text{kN / m}$$

Add water pressure:

$$F_{hW} = 355 + 10\frac{3^2}{2} \text{ kN / m}$$
$$= 400 \text{ kN/m}$$

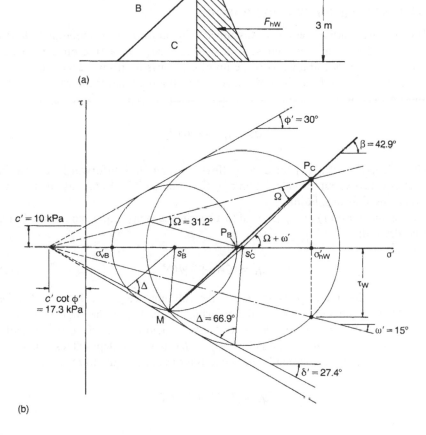

(a)

(b)

Figure 7.5 Example 7.2.

The angle b between the direction of the discontinuity and the plane of action of the major principal stress in the lower stressed zone (zone B) is given by equation 7.13:

$$b = \tfrac{1}{2}(\Delta + \delta')$$
$$= \tfrac{1}{2}(66.9° + 27.4°)$$
$$= 47.1° \text{ to the vertical}$$

and the angle β to the horizontal is 42.9°.

7.6 Smooth foundation on cohesionless soil (ϕ', $c' = 0$)

7.6.1 Bearing capacity expression

The ultimate bearing pressure q_F (assumed to be uniform) of a smooth foundation on the surface of a cohesionless soil, with surcharge p_S acting on the surface adjacent to the loaded area, is made up of $q_F(1)$ arising from the surcharge and $q_F(2)$ arising from the weight of the soil. It is assumed that these can be superimposed, giving

$$q_F = q_F(1) + q_F(2) \tag{7.21}$$

The calculation of $q_F(1)$ for a weightless soil is a straightforward exercise in plasticity theory, but the calculation of $q_F(2)$ is more difficult, because there is no direct closed form solution.

Solutions for $q_F(1)$ and $q_F(2)$ lead to a bearing capacity expression of the form, for a dry soil:

$$q_F = p_S N_q + \tfrac{1}{2}\gamma B N_\gamma \tag{7.22}$$

where p_S is the surcharge pressure, γ is the bulk unit weight of the soil, B is the foundation width and N_q, N_γ are factors which depend upon the angle of friction ϕ'. If a foundation is embedded at a depth D_e below the surface it is usually assumed, for a shallow foundation ($D_e \leqslant B$), that this depth does not itself contribute to the shear strength, but it exerts a surcharge equal to D_e, thus

$$q_F = \gamma D_e N_q + \tfrac{1}{2}\gamma B N_\gamma \tag{7.23}$$

7.6.2 Evaluation of N_q

Values of N_q can be obtained using simple plasticity theory for a weightless soil. The method adopted in this Section is based on the introduction of stress discontinuities. An alternative approach using stress characteristics is given in Section 8.3.1.

(a) Single discontinuity

A single vertical stress discontinuity can be inserted below the corner of the applied loading $q_F(1)$, separating zones C and B as shown in Figure 7.6(a). As the principal stresses are vertical and horizontal in both these zones, there is no friction acting tangentially to the discontinuity, and the stress circles in the two zones are as shown in Figure 7.6(b). It can be seen from Figure 7.6(b) that

$$\frac{q_F(1)}{p_S} = N_q = \left(\frac{1+\sin\phi'}{1-\sin\phi'}\right)^2 \tag{7.24}$$

(b) Two or more discontinuities

Two discontinuities can be inserted as shown in Figure 7.7(a), creating a zone D separating zones C and B. As the principal stress directions in zones C and B differ by 90°, the rotation in stress direction across each discontinuity is 45°, if the rotation is divided equally between the two. The resulting stress circles are shown in Figure 7.7(b).

From Equation 7.15:

$$90° - \Delta = 45°$$
$$\therefore \quad \Delta = 45°$$

From Figure 7.7(b):

$$q_F(1) = s_C'(1+\sin\phi')$$
$$p_S = s_C'(1-\sin\phi')$$
$$\frac{q_F(1)}{p_S} = N_q = \frac{s_C'(1+\sin\phi')}{s_B'(1-\sin\phi')} \tag{7.25}$$

Equation 7.17:

$$\frac{s_C'}{s_D'} = \frac{s_D'}{s_D'} = \frac{\sin(\Delta+\delta')}{\sin(\wedge-\delta')}$$

$$\therefore \frac{s_C'}{s_B'} = \left(\frac{\sin(\Delta+\delta')}{\sin(\Delta-\delta')}\right) \tag{7.26}$$

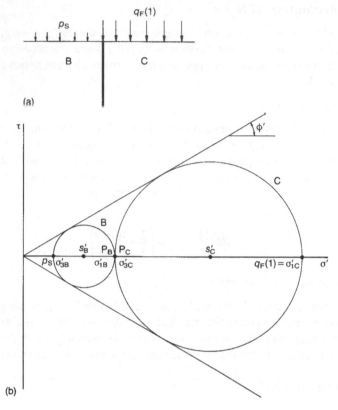

Figure 7.6 Smooth foundation: (a) single discontinuity; (b) corresponding effective stress circles for weightless soil.

As ϕ' and Δ are both known, δ' can be found from equation 7.12. Combining equations 7.25 and 7.26 gives

$$N_q = \left(\frac{\sin(\Delta + \delta')}{\sin(\Delta - \delta')} \right)^2 \left(\frac{\sin(1 + \phi')}{\sin(1 - \phi')} \right) \tag{7.27}$$

The angles β_{CD} and β_{BD} of the discontinuities can be found from either Figure 7.7(b) or equations 7.13 and 7.14. For $\phi' = 30°$, $\Delta = 45°$, the value of $\delta' = 20.7°$ is given by equation 7.12. Substituting these values of Δ and δ' into equation 7.13 gives $b = 32.9°$, which is the angle between discontinuity separating zones B and D and the plane on which the major principal stress acts in the lower stressed zone B, which is vertical. Thus

$$\beta_{BD} = 90° - 32.9° = 57.1°$$

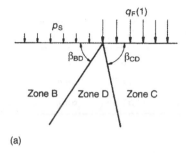

(a)

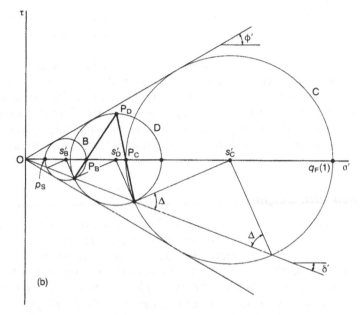

(b)

Figure 7.7 Smooth foundation: (a) two discontinuities; (b) corresponding effective stress circles for weightless soil.

The angle β_{CD} between the discontinuity separating zones C and D and the horizontal plane in zone C, on which the major principal stress acts, is given by equation 7.14. Thus

$$\beta_{CD} = 77.8°$$

The procedure above for two discontinuities can be followed for any number of discontinuities n, for which equation 7.27 becomes

$$N_q = \left(\frac{\sin(\Delta + \delta')}{\sin(\Delta - \delta')}\right)^n \left(\frac{(1 + \sin\phi')}{(1 - \sin\phi')}\right) \qquad (7.28)$$

Table 7.1 Values of N_q

ϕ'	$n = 1$	$n = 2$	$n = 5$	$n = \infty$
20°	4.2	5.6	6.3	6.4
30°	9.0	14.7	17.7	18.4
40°	21.2	43.6	59.8	64.2

A solution for an infinite number of discontinuities can be found by putting $\sin\phi' \rightarrow \phi'$ as $\phi' \rightarrow 0$. This is discussed more fully in Section 8.3.1, as the stress characteristics solution corresponds to that for an infinite number of discontinuities.

Values of N_q for $\phi' = 20°$, 30° and 40° given by equation 7.28 for $n = 1$, 2 and 5 discontinuities are given in Table 7.1, together with N_q for $n = \infty$ obtained as described in Section 8.3.1. It can be seen in Table 7.1 that a single discontinuity gives poor results; while considerably improved by the use of two discontinuities, results are still unsatisfactory. The use of five discontinuities gives results reasonably comparable to those given by the more refined approach using stress characteristics.

7.6.3 Soil with weight

The assumption of a weightless soil is clearly impractical in the case of a foundation on the surface of a cohesionless soil with no surcharge, as this will give a lower bound solution of $q_F = 0$. In practice foundations are usually embedded at some depth D_e and the bearing capacity is calculated on the assumption that the depth D_e of soil does not contribute to the shear strength, but acts as a surcharge γD_e. For shallow foundations where $D_e \leqslant B$, the foundation width, the weight of the soil below foundation level will make a significant contribution to the bearing capacity.

Foundation loading on the surface of a cohesionless soil with weight has been considered by Sokolovski (1965), who presents a solution for a smooth foundation obtained by the numerical integration of differential equations associated with slip lines. A simple solution using stress discontinuities is not available, but the crude procedure illustrated in Figure 7.8 can be implemented to provide approximate values of N_γ.

The vertical line AB below the corner of the loaded area in Figure 7.8 can be regarded as a 'retaining wall' separating the 'active' zone below the loaded area from the 'passive' zone outside the loaded area. Although it is not possible to match the stresses across AB it is possible to equate the active and passive forces over a specified depth D_s.

It can be seen from the active and passive stress distributions in Figures 7.8(b) and (c) that the horizontal active and passive forces F_{hA} and F_{hP} over depth D_s are given by

$$F_{hA} = K_A[q_F(2)D_S + \tfrac{1}{2}\gamma D_s^2 \qquad (7.29)$$

$$F_{hA} = K_p(\tfrac{1}{2}\gamma D_s^2) \qquad (7.30)$$

where K_A, K_P are the active and passive coefficients of lateral stress.
   Equating equations 7.29 and 7.30 gives

$$q_F(2) = \tfrac{1}{2}\gamma D_s\left(\frac{K_P}{K_A} - 1\right) \qquad (7.31)$$

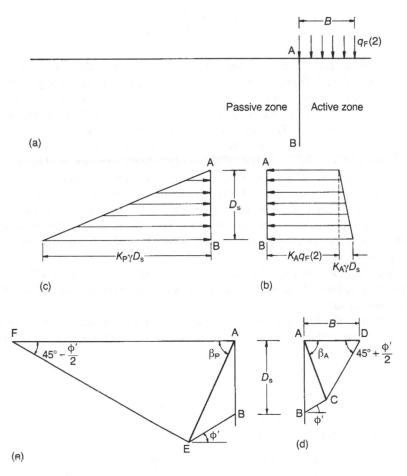

Figure 7.8 Smooth foundation on a soil with weight: (a) active and passive zones; (b,c) active and passive stresses; (d,e) assumed active and passive blocks.

Assuming D_s to be a linear function of foundation width B, i.e.

$$D_s = mB \tag{7.32}$$

gives

$$q_F(2) = \tfrac{1}{2} Bm \left(\frac{K_P}{K_A} - 1 \right) \tag{7.33}$$

which is the correct form of expression for $q_F(2)$, in which

$$N_\gamma = m \left(\frac{K_P}{K_A} - 1 \right) \tag{7.34}$$

Values of K_A and K_P can be found using the procedures outlined in Sections 7.4 and 7.5. If the full friction angle $\omega' = \phi'$ is assumed to act tangentially along AB, the values of K_A and K_P for $\phi' = 20°$, $30°$ and $40°$ are those shown in Table 7.2.

In order to calculate the depth D_s, a reasonable approach is to assume that the 'active' block ABCD is bounded by the lines BC, CD shown in Figure 7.8(d), which are lines of maximum stress obliquity. Lines BC, CD can easily be shown to have angles to the horizontal of ϕ' and $(45° + \tfrac{1}{2}\phi')$ respectively. The 'passive' block can also be assumed to be bounded by maximum stress obliquity lines BE, EF in Figure 7.8(e), although the geometry of these is not needed in the calculations. Point C can be fixed by the angle of the discontinuity β_A in the 'active' zone, values of which can be determined by procedures described in Section 7.4. These values are given in Table 7.2 for $\phi' = 20°$, $30°$ and $40°$, together with values of $D_s/B = m$.

Inserting the values of K_A, K_P and m given in Table 7.2 into equation 7.34 yields the magnitudes of N_γ shown in Table 7.3. Values derived by Sokolovski (1965) are also shown.

Although the differences in N_γ in Table 7.3 are quite large, they may not have such a substantial influence on calculated bearing capacity. The largest percentage

Table 7.2 Values of K_A, K_P, β_A and $D_s/B = m$

ϕ'	K_A	K_P	β_A	$D_s/B = m$
20°	0.434	2.27	70.6°	1.07
30°	0.238	3.93	72.8°	1.33
40°	0.179	7.73	75.3°	1.68

Table 7.3 Values of N_γ

	N_γ	
ϕ'	Eq. 7.34	Sokolovski
20°	4.5	3.2
30°	17	15
40°	71	86

difference is for $\phi' = 20°$, which normally denotes a clay and hence bearing capacity is more likely to be based on undrained rather than drained strength parameters.

Taking a simple case of a foundation of 1 m width at a depth of 0.75 m in a dry soil for which $\gamma = 20$ kN/m³ and $\phi' = 40°$, the calculated ultimate bearing capacity q_F (using N_q as given by the stress characteristic solution in Section 8.3.1) is

$$\text{Equation 7.34: } N_\gamma = 71 \quad q_F = 1.67 \text{ MPa}$$

$$\text{Sokolovski: } N_\gamma = 86 \quad q_F = 1.82 \text{ MPa}$$

These differences in calculated bearing capacity are of the same magnitude as the differences caused by an error of 0.5° in the assumed value of ϕ'. For example, using the Sokolovski parameters, the calculated bearing capacity for $\phi' = 39.5°$ drops to 1.68 MPa.

Designers would of course be wary of assuming ultimate bearing capacities of the magnitudes calculated above, having in mind the large effect of a small difference in ϕ' from that assumed in the calculation. In addition, large strains may be required to generate fully the lateral passive resistance, leading to substantial settlements or even punching failure at foundation pressures well below the calculated ultimate bearing capacity.

7.7 Silo problem

Silos containing granular materials have been observed to fail by bursting of the silo wall during flow, at the level where the lower inclined portion of the wall meets the vertical wall. This is a point of strong stress concentration, as shown below.

7.7.1 Smooth walls ($c' = 0$, ϕ' material)

Consider a silo with the walls of the lower Section inclined at 45°, as shown in Figure 7.9(a), with a horizontal pressure $\sigma'_{nW}(B)$ on the vertical portion of the wall immediately above the change in slope. The change in direction of major principal stress for smooth walls is 45°.

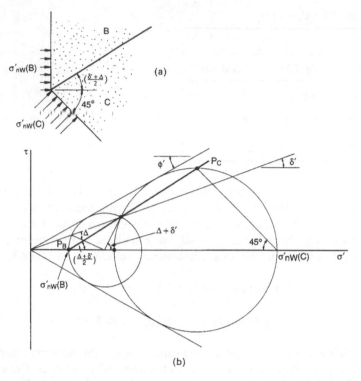

Figure 7.9 Silo with smooth walls, single discontinuity solution: (a) wall stresses;
(b) effective stress circles.

(a) Single discontinuity

$$\text{Equation 7.15: } 90° - \Delta = 45°$$

$$\therefore \quad \Delta = 45°$$

$$\text{Equation 7.12: } \sin \delta' = \sin \Delta \sin \phi'$$

$$= 0.707 \sin \phi'$$

$$\text{For } \phi' = 30°, \delta' = 20.7°$$

$$\text{Equation 7.18: } \frac{\sigma'_{nw}(C)}{\sigma'_{nw}(B)} = \left[\frac{\sin(\Delta + \delta')}{\sin(\Delta - \delta')} \right] \left[\frac{1 + \sin \phi'}{1 - \sin \phi'} \right]$$

$$= 6.64$$

$$\text{For } \phi' = 40°, \delta' = 27.03°$$

$$\frac{\sigma'_{nw}(C)}{\sigma'_{nw}(B)} = 14.2$$

(b) Two discontinuities

$$\text{Equation 7.15: } (90 - \Delta) = 22.5°$$

$$\therefore \quad \Delta = 67.5°$$

$$\sin \delta' = 0.924 \sin \phi'$$

$$\text{For } \phi' = 30°, \ \delta' = 27.5'$$

$$\text{Equations 7.18, 7.26: } \frac{\sigma'_{nW}(C)}{\sigma'_{nW}(B)} = \left[\frac{\sin(\Delta + \delta')}{\sin(\Delta - \delta')} \right]^2 \left[\frac{1 + \sin \phi'}{1 - \sin \phi'} \right]$$

$$= 7.20$$

$$\text{For } \phi' = 40°, \ \delta' = 36.4°$$

$$\frac{\sigma'_{nW}(C)}{\sigma'_{nW}(B)} = 16.2$$

Table 7.4 gives the values of $\sigma'_{nW}(C)/\sigma'_{nW}(B)$ for (a) a single discontinuity, (b) two discontinuities. The magnitudes for infinite numbers of discontinuities (c) are also given (see Section 8.3.3).

7.7.2 Rough walls (c' = 0, φ' soil)

Consider again a silo with the walls of the lower Section inclined at 45°, as shown in Figure 7.10, with horizontal pressure $\sigma'_{nW}(B)$ on the vertical portion of the wall immediately above the change in slope. Assume the friction angle between the wall and the flowing material is equal to $\phi'/2$. For $\phi' = 30°$, $\omega' = 15°$,

$$\text{Equation 7.19: } \sin \omega' = \sin \phi' \sin \Omega$$

$$\therefore \quad \Omega = 31.2°$$

and

$$\Omega - \omega' = 16.2°$$

$$\Omega + \omega' = 46.2°$$

Table 7.4 Values of $\sigma'_{nW}(C)/\sigma'_{nW}(B)$

	$\sigma'_{nW}(C)/\sigma'_{nW}(B)$		
ϕ'	(a) n = 1	(b) n = 2	(c) n = ∞
30°	6.6	7.2	7.4
40°	14.2	16.2	17.2

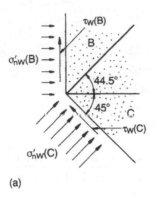

(a)

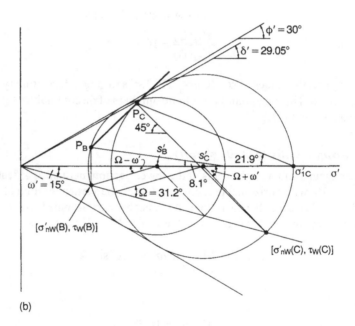

(b)

Figure 7.10 Silo with rough walls, single discontinuity solution: (a) wall stresses; (b) effective stress circles.

Thus, from Figure 7.10(b) the major principal stress is inclined at 8.1° to the vertical in zone B and at 21.9° in zone C. The change in direction of major principal stress from zone B to zone C is 13.8°.

(a) Single discontinuity

$$\text{Equation 7.15: } 90° - \Delta = 13.8°$$
$$\Delta = 76.2°$$

Equation 7.12: $\sin \delta' = \sin \phi' \sin \Delta$
$$\therefore \delta' = 29.05°$$

Equation 7.17: $\dfrac{s'_C}{s'_B} = \dfrac{\sin(76.2° + 29.05°)}{\sin(76.2° - 29.05°)}$
$$= 1.316$$

From Figure 7.10(b):

$$s'_B = \sigma'_{nW}(B)[1 + \tan\omega'\cot(\Omega - \omega')] \qquad (7.35)$$

$$\sigma'_{nW}(C) = s'_C[1 + \sin\phi'\cos(\Omega - \omega')] \qquad (7.36)$$

Combining equations 7.35 and 7.36 and putting $\Omega = 31.2°$, $\omega' = 15°$ gives

$$\frac{\sigma'_{nW}(C)}{\sigma'_{nW}(B)} = 3.40$$

From equation 7.13 the angle between the discontinuity and the plane on which the major principal stress acts in the lower stressed zone is given by $\frac{1}{2}(\Delta + \delta') = 52.6°$. As the principal stress in zone B acts on a plane at 8.1° to the horizontal, the angle β of the discontinuity to the horizontal is 44.5°.

(b) Two discontinuities

Equation 7.15, for each discontinuity: $90° - \Delta = 6.9°$
$$\therefore \Delta = 83.1°$$

Equation 7.12: $\delta' = 29.9°$

Equation 7.26: $\dfrac{s'_C}{s'_B} = \dfrac{\sin^2(83.1° + 29.9°)}{\sin^2(83.1° - 29.9°)} = 1.320$

Applying this value of s'_C/s'_B to equations 7.35 and 7.36 gives

$$\frac{\sigma'_{nW}(C)}{\sigma'_{nW}(B)} = 3.42$$

Thus, the assumption of two or more discontinuities has only a slight influence on the result because of the small change in direction of the major principal stress from zone B to zone C. (See also Section 8.3.3.)

It can be seen that the effect of the rough walls is to reduce the magnitude of $\sigma'_{nW}(C)/\sigma'_{nW}(B)$. If $\phi' = 30°$, and two discontinuities are assumed, the ratio for ω' = 15° is 3.4 compared to 7.2 for $\omega' = 0$.

EXAMPLE 7.3 STRESS CONCENTRATION IN SILO WALL

In an effort to reduce stress concentration at the junction of the vertical walls and lower inclined Section of a silo, but at the same time facilitate flow of the granular material in the silo, an experiment is conducted with a silo keeping the vertical walls rough and making the inclined surface very smooth. If the inclined surface has an angle of 50° to the horizontal, as shown in Figure 7.11(a), find the relationship between the normal stresses on the wall and inclined surface immediately above and below the junction if the friction angle is $0.75\phi'$ between the granular material and vertical walls and $0.25\phi'$ between the granular material and the inclined surface. Assume $\phi' = 32°$. Obtain the solution using two discontinuities. Find the angles of the discontinuities.

Solution

The angles of planes on which the major principal stresses act in zones B and C can be found as shown in Figure 7.11(b) for zone B and in Figure 7.11(c) for zone C (not to scale). The stress points $\sigma'_{nW}(B)$ and $\sigma'_{nW}(C)$ are found by projecting lines at angles of $\omega' = 24°$ and $8°$ respectively in zones B and C. As the shear stresses on soil element faces parallel to the walls are clockwise, the stress points plot below the $\tau = 0$ line. Once the stress points are located, the pole points for planes P_B and P_C can be found, and from these lines through σ'_{1B} and σ'_{1C} give the directions of planes on which σ'_{1B} and σ'_{1C} act.

In zone B:

$$\text{Equation 7.19:} \quad \sin\Omega_B = \frac{\sin 24°}{\sin 32°}$$

$$\therefore \quad \Omega = 50°$$

From Figure 7.11(b) the angle to the horizontal of the plane on which σ'_{1B} acts is

$$\tfrac{1}{2}(\Omega_B - \omega'_B) = 13°$$

In zone C:

$$\text{Equation 7.19: } \sin\Omega_C = \frac{\sin 8°}{\sin 32°}$$

$$\therefore \quad \Omega_C = 15.2°$$

From Figure 7.11(c) the angle to the horizontal of the plane on which σ'_C acts is

$$50° - \tfrac{1}{2}(\Omega_C + \omega') = 38.4°$$

Therefore the change in direction of σ'_1 from zone B to zone C is 25.4°. Assuming two discontinuities, the change across each discontinuity is 12.7°. For each discontinuity:

$$\text{Equation 7.15: } 90° - \Delta = \theta$$

$$\therefore \quad \Delta = 77.3°$$

$$\text{Equation 7.12: } \sin\delta' = \sin 32° \sin 77.3°$$

$$\therefore \quad \delta' = 31.1°$$

$$\text{Equation 7.26: } \frac{s'_C}{s'_B} = \left(\frac{\sin(77.3° + 31.1°)}{\sin(77.3° - 31.1°)}\right)^2 = 1.728$$

$$\text{Equation 7.17: } \frac{s'_D}{s'_B} = \frac{s'_C}{s'_D} = \frac{\sin(77.3° + 31.1°)}{\sin(77.3° - 31.1°)}$$

The three circles for zones B, D and C can now be drawn with centres s'_B, s'_D and s'_C, as shown in Figure 7.11(d).

From equations 7.35 and 7.36 (or Figures 7.11(b) and (c)):

$$s'_B = \sigma'_{nW}(B)(1 + \tan 24° \cot 26°)$$

$$= 1.91\sigma'_{nW}(B)$$

$$\sigma'_{nW}(C) = s'_C(1 + \sin 32° \cos 23.2°)$$

$$= 1.49 s'_C$$

$$\therefore \frac{\sigma'_{nW}(C)}{\sigma'_{nW}(B)} = 1.91 \times 1.49 \times \frac{s'_C}{s'_B} = 4.91$$

The orientation of the discontinuity between zones B and D is given by equation 7.13:

$$b = \tfrac{1}{2}(77.3° + 31.1°)$$

$$= 54.2°$$

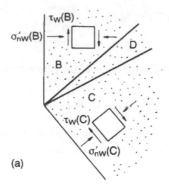

(a)

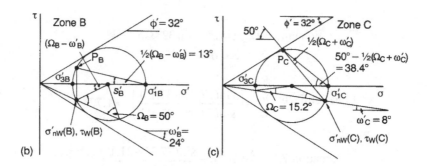

(b)   (c)

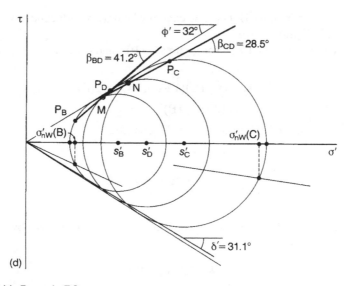

(d)

Figure 7.11 Example 7.3.

This is the angle between the discontinuity and the plane on which σ'_{1B} acts. Thus the angle β_{BD} to the horizontal is

$$\beta_{BD} = 54.2° - 13° = 41.2°$$

The orientation of the discontinuity between zones D and C is given by equation 7.14:

$$c = \tfrac{1}{2}(180° - 77.3° + 31.1°)$$
$$= 66.9°$$

This is the angle between the discontinuity and the plane of action on which σ'_{1C} acts. Thus the angle β_{CD} to the horizontal is

$$\beta_{CD} = 66.9° - 38.4°$$
$$= 28.5°$$

The angles β_{BD}, β_{CD} can also be obtained from Figure 7.11(d), by a line through P_B and the common stress point M for circles B and D to obtain β_{BD}, and by a line through P_C and the common stress point N for circles C and D to obtain β_{CD}. These lines intersect on circle D at the pole point P_D.

Chapter 8

Stress characteristics and slip lines

8.1 Stress characteristics

Where the geometry of a problem makes their use possible, stress character-istics should give a better lower bound solution than the assumption of stress discontinuities, and in some very simple cases can give an exact solution. Stress characteristics may be straight or curved and must obey certain restraints. These restraints and the full theory of stress characteristics have been the subject of many publications (e.g. Hencky, 1923; Sokolovski, 1965; Abbot, 1966; Houlsby and Wroth, 1982; Atkinson, 1981). It is not intended here to repeat the full theory, but simply to show that Mohr circles can be used to obtain expressions for stress change along a characteristic by the summation of infinitesimally small stress discontinuities.

Abbot (1966) introduces the concept of a characteristic as a propagation path; that is a path followed by some entity, such as a geographical form or a physical disturbance, when that entity is propagated. In a two-dimensional system the characteristics appear as lines on a physical surface. In a loaded body the entity can be a defined stress condition, and in a plastically deforming soil the concept is most usefully employed in tracing the propagation paths of the maximum shear stresses in undrained deformation, or the Mohr–Coulomb failure condition in drained deformation.

As seen in Sections 6.3 and 7.3, a change in stress direction can occur across a discontinuity. If the discontinuities are closely spaced the stress path approaches a smooth curve as the spacing of the discontinuities approaches zero.

8.2 Undrained stress characteristics

The two-dimensional stresses at a point in a mass of saturated clay experiencing shear failure under undrained conditions are shown in Figure 8.1(a), and may be represented by the Mohr stress circle in Figure 8.1(b). As the soil is at failure the stress circle touches the c_u envelopes. The pole point for planes P is easily found by a line through σ_1 in Figure 8.1(b) parallel to the direction of the plane on which σ_1 acts in Figure 8.1(a). Two lines can be drawn through P and the points where

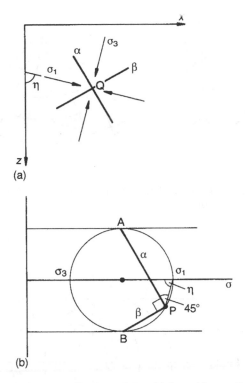

Figure 8.1 α and β stress characteristics in undrained failure: (a) two-dimensional stress system; (b) total stress circle.

the stress circle touches the stress envelopes $+c_u$ and $-c_u$. The line through P and $+c_u$ is the direction of the α characteristic at the stress point in Figure 8.1(a) and the line through P and is the direction of the β characteristic. It is clear from Figure 8.1(b) that these lines, and hence the α and β characteristics at their points of intersection, are always orthogonal for undrained total stress representation. It also follows from the above that shear stresses along an α characteristic are always positive, or anticlockwise, and along a β characteristic they are always negative, or clockwise.

The simplest sets of characteristics which intersect orthogonally are parallel straight lines, as shown in Figure 8.2(a), and circular arcs crossing straight lines radiating from a centre, as shown in Figure 8.2(b). As shown below, a number of simple problems can be addressed using these two basic sets of characteristics, but more complex problems such as cone penetration (Houlsby and Wroth, 1982) require the generation of more complex sets of characteristics.

Referring to equation 6.8 and Figure 8.3 it can be seen that a small change in stress state across a discontinuity (noting $\sin d\theta \to d\theta$ as $d\theta \to 0$ is given by

$$ds = 2c_u\, d\theta \qquad (8.1)$$

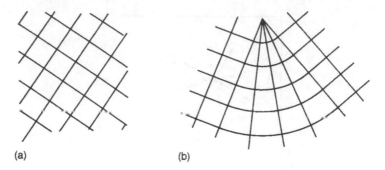

(a)   (b)

Figure 8.2 Orthogonal characteristics: (a) parallel straight lines; (b) radiating straight lines and circular arcs.

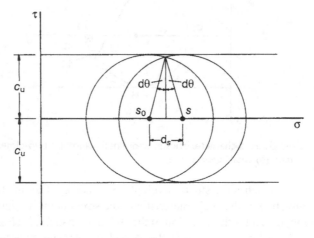

Figure 8.3 Small stress change across a discontinuity.

That is

$$ds = 2(+c_u)d\theta \text{ along an } \alpha \text{ characteristic} \tag{8.2a}$$

$$ds = 2(-c_u)d\theta \text{ along a } \beta \text{ characteristic} \tag{8.2b}$$

where $d\theta$ is the change in stress direction, and hence the change in the direction of the characteristic, across the discontinuity. The negative sign for the β characteristic is essentially a mathematical device to distinguish the characteristics, and does not necessarily indicate a reduction in the magnitude of stress.

The change in stress along a finite length of characteristic can be found by integrating equations 8.2a and 8.2b to give

$$s - s_0 = \pm 2c_u (\theta - \theta_0) \qquad (8.3)$$

This equation ignores body forces, which can be superimposed on equation 8.3 to give

$$s - s_0 = \pm 2c_u (\theta - \theta_0) + \gamma(z - z_0) \qquad (8.4)$$

where γ is the unit weight of the soil and z is depth.

It follows from equation 8.3 that for a weightless material, which is often assumed in solving plasticity problems, there is no change in stress (that is, the stress state is constant) along a linear stress characteristic.

A number of other corollaries follow from Hencky's (1923) rule which can be explained by consideration of Figure 8.4, in which the characteristics make up a curvilinear rectangle JKLM. The short broken lines which bisect the inter-

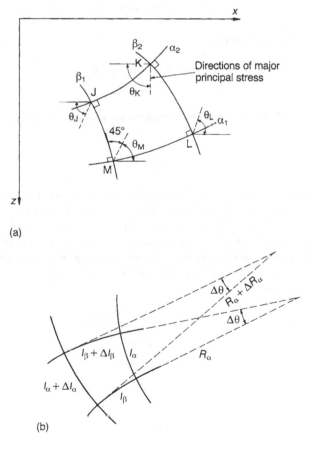

(a)

(b)

Figure 8.4 Hencky diagram.

Section angles give the directions of the major principal stresses, and the rotations of the characteristics can be represented by the rotations of the principal stress directions. Thus, applying equation 8.4 progressively around the curvilinear rectangle gives:

1. along α characteristics,

$$J \rightarrow K \qquad\qquad s_K - s_J = 2c_u(\theta_K - \theta_J) + \gamma(z_K - z_J) \qquad\qquad (8.5a)$$

$$L \rightarrow M \qquad\qquad s_M - s_L = 2c_u(\theta_M - \theta_L) + \gamma(z_M - z_L) \qquad\qquad (8.5b)$$

2. along β characteristics,

$$K \rightarrow L \qquad\qquad s_L - s_K = -2c_u(\theta_L - \theta_K) + \gamma(z_L - z_K) \qquad\qquad (8.5c)$$

$$M \rightarrow J \qquad\qquad s_J - s_M = -2c_u(\theta_J - \theta_M) + \gamma(z_J - z_M) \qquad\qquad (8.5d)$$

Summing these equations gives

$$\theta_K + \theta_M - \theta_L - \theta_J = 0 \qquad\qquad (8.6)$$

or

$$\theta_K - \theta_J = \theta_L - \theta_M \qquad\qquad (8.7a)$$

$$\theta_K - \theta_L = \theta_J - \theta_M \qquad\qquad (8.7b)$$

These equations are an expression of Hencky's rule that the angle subtended by neighbouring lines of one family, where they intersect a line of the second family, is constant. In setting up fields of characteristics it is therefore convenient to fix the intersection nodes to give a selected constant amount of rotation, say 15° or 0.262 radians. With this amount of rotation it can be seen from equation 8.4 that the change of s from node to node will be $0.524c_u + \gamma z$.

From Figure 8.4:

$$l_\alpha + \Delta l_\alpha = (R_\alpha + \Delta R_\alpha)\Delta\theta$$
$$l_\alpha = R_\alpha \Delta\theta$$
$$\therefore \quad \Delta l_\alpha = \Delta R_\alpha \Delta\theta \qquad\qquad (8.8)$$

But also

$$\Delta l_\alpha = l_\beta \Delta\theta$$
$$\therefore \quad \Delta R_\alpha = l_\beta \qquad\qquad (8.9)$$

It follows that:

1. in travelling along a characteristic, the radius of curvature of each characteristic of the other family must change by amounts equal to the distance between the intersections;
2. if any one line in a family is straight, all lines in that family must be straight, and the lines in the other family must be circular arcs with a common centre (which may be at infinity, giving sets of parallel straight lines).

In order to obtain simple indicative solutions to soil mechanics problems it is common to assume the soil to be weightless, which allows the use of straight line and circular arc characteristics. The introduction of weight to the soil causes the characteristics to be curved.

EXAMPLE 8.1 ACTIVE FORCE ON RETAINING WALL DUE TO SURCHARGE

Solve Example 6.1(b) using stress characteristics.

Solution

Close to the wall the major principal stress acts on a plane at $32°$ to the horizontal, as shown in Figure 6.4(d), and there is thus a rotation of $32°$ (0.56 radians) in the major principal stress direction.

$$\text{Equation 8.3: } s_C - s_B = 2 \times 0.56 c_u$$
$$= 22.4 \text{ kPa}$$

Referring to Figures 6.4(d) and (e):

$$q_s = 80 = s_C + 20 \text{ kPa}$$
$$\sigma_{hw} = s_B - c_u \cos 64°$$
$$= 80 - 22.4 - 20 - 20 \cos 64° \text{ kPa}$$
$$\therefore \quad \sigma_{hw} = 28.8 \text{ kPa}$$
$$\therefore \quad F_{hw} = 86.4 \text{ kN/m}$$

This value of F'_{hw} compares with $F_{hw} = 87.9$ kN/m obtained using a single discontinuity solution.

The stress characteristics are shown in Figure 8.5.

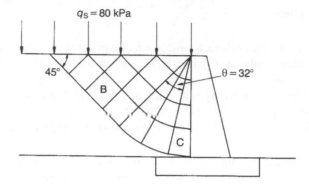

Figure 8.5 Example 8.1.

8.2.1 Smooth strip footing

Consider a uniformly loaded strip footing as shown in Figure 8.6(a) at the point of failure under an applied pressure q_F. The shear stress on the surface is zero and the soil is weightless. A uniform surcharge loading p_S acts on the surface adjacent to the footing.

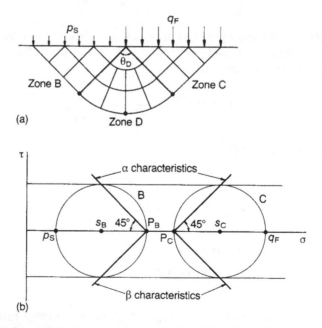

Figure 8.6 Smooth strip footing under vertical loading: (a) undrained stress characteristics; (b) total stress circles.

It can be seen in Figure 8.6(b) that the stress characteristics in zone B (passive zone) and zone C (active zone) act at 45° to the horizontal. The directions of the major principal stresses are, respectively, horizontal and vertical. Consequently the angle θ_D of the fan zone is 90° or $\pi/2$ radians. The curved lines in zone D are α lines, as shear stresses along them are anticlockwise.

From equation 8.3:

$$s_C - s_B = 2c_u \frac{\pi}{2} = \pi c_u \qquad (8.10)$$

Thus, from the geometry of Figure 8.6(b):

$$q_F = c_u (2 + \pi) + p_S$$

i.e.

$$q_F = 5.14c_u + p_S \qquad (8.11)$$

In order to present this as a true lower bound solution it is strictly necessary to demonstrate that every element of soil outside the plastic zones is in stress equilibrium and not violating the yield conditions. This can be shown (Bishop, 1953) and in fact this is both a lower bound solution and a correct solution.

EXAMPLE 8.2 BEARING CAPACITY UNDER STRIP LOADING

Solve Example 6.2 using stress characteristics.

Solution

Case (a): as illustrated in Figure 6.8(c), the principal stress rotates by 70° = 1.22 radians from zone C to zone B.

$$\text{Equation 8.3: } s_C - s_B = 2 \times 1.22c_u$$
$$= 2.44c_u$$

$$\therefore q_F = (2 + 2.44)c_u$$
$$= 4.44c_u$$
$$= 444 \text{ kPa}$$

This value of $q_F = 444$ kPa compares with $q_\Gamma = 429$ kPa obtained using two discontinuities.

Similarly for case (b) it can be shown that the stress characteristic solution gives $q_F = 584$ kPa compared with $q_F = 528$ kPa obtained using two discontinuities.

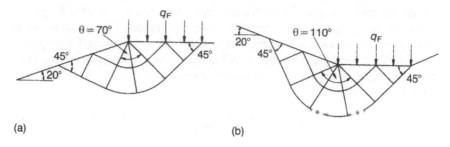

Figure 8.7 Example 8.2.

The stress characteristics are shown in Figure 8.7.

8.2.2 Strip footing under vertical and horizontal loading

Consider the strip footing in Figure 8.8(a) which applies a shear stress τ_h along the surface and q_{vF}, which causes plastic failure, normal to the surface. The soil is assumed weightless.

The directions of the characteristics in zones B and C are indicated in Figure 8.8(b).

In zone B they are at 45° to the horizontal. It can be seen from the expanded Section of Figure 8.8(b), shown in Figure 8.8(c), that

$$\frac{\tau_h}{\sin\theta_C} + \frac{c_u}{\sin\theta_C} = 2c_u \sin\theta_C$$

$$\therefore \quad \tau_h = c_u(2\sin^2\theta_C - 1) \tag{8.12}$$

Also, from Figures 8.8(b) and 8.8(d):

$$\theta_D = \pi - \theta_C - \frac{\pi}{4} = \frac{3\pi}{4} - \theta_C \tag{8.13}$$

Equation 6.8:

$$s_C - s_B = 2c_u\left(\frac{3\pi}{4} - \theta_C\right) \tag{8.14}$$

Thus from Figures 8.8(b) and 8.8(d):

$$q_{vF} = p_S + c_u + 2c_u\left(\frac{3\pi}{4} - \theta_C\right) + c_u \sin 2\theta_C$$

$$\therefore \quad q_{vF} = p_S + c_u\left(1 + \frac{3\pi}{2} - 2\theta_C + \sin 2\theta_C\right) \qquad (8.15)$$

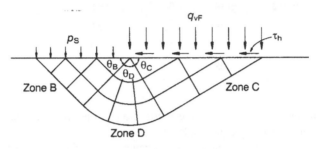

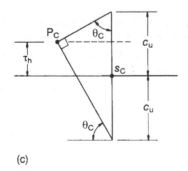

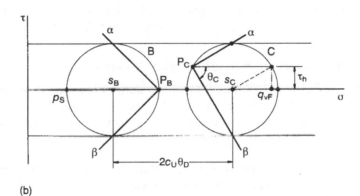

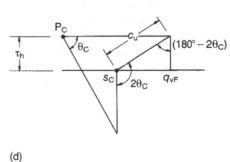

Figure 8.8 Strip footing under vertical and horizontal loading: (a) undrained stress characteristics; (b) total stress circles; (c,d) expanded Sections of stress diagram.

Table 8.1 Calculated values of strip footing loading parameters

$\dfrac{\tau_h}{c_u}$	θ_C (rad)	θ_D (rad)	$\dfrac{q_{vF} - p_S}{c_u}$	$\dfrac{q_{vF} - p_S}{\tau_h}$
0	0.785	1.571	5.14	∞
0.25	0.912	1.444	4.86	19.5
0.50	1.047	1.309	4.48	9.0
0.75	1.209	1.146	3.95	5.3
1.0	1.571	0.785	2.57	2.6

The relationship between θ_C and the inclination of the applied load is found by dividing equation 8.15 by equation 8.12.

It is more straightforward and useful, however, to calculate values of $(q_{vF} - p_s)/c_u$ for specific values of τ_h/c_u. For each assumed value of τ_h/c_u, θ_C can be calculated from equation 8.12, θ_D from equation 8.13 and, thus, $(q_{vF} - p_S)/c_u$ from equation 8.15. Calculated values are given in Table 8.1, together with $(q_{vF} - p_S)/\tau_h$.

8.2.3 Flow between rough parallel platens

The stress characteristics for the problem shown in Figure 6.9, assuming the full undrained shear strength is developed between the soil and the platens, are shown in Figure 8.9(a). These can be constructed following the procedure outlined below and satisfying the stated requirements:

1. At all points such as B, F, J and M along the centre line of the deforming soil the vertical and horizontal shear stresses must be zero to satisfy symmetry, and consequently all stress characteristics must cross this line at an angle of 45° to the horizontal.
2. The horizontal stress σ_s acting on the free face ABC is a principal stress, and thus the stress characteristics meeting this face must do so at an angle of 45° to the horizontal as shown in Figure 8.9(b). It follows from this and requirement 1 above that the stress characteristics within the wedge ABCF are all linear and at 45° to the horizontal. It is thus also a zone of constant stress, in which

$$\sigma_v = \sigma_s + 2c_u \tag{8.16}$$

$$s = \sigma_s + c_u \tag{8.17}$$

3. At all points along the interface between the soil and the platens a horizontal shear stress $\tau_h = c_u$ acts, as well as a vertical stress σ_v. As seen in the stress

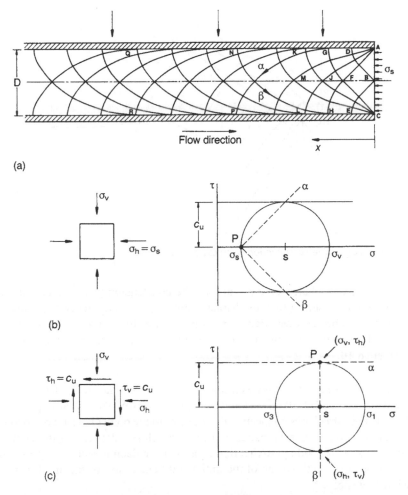

Figure 8.9 Undrained flow between rough parallel platens: (a) stress characteristics; (b) total stress circle at free face; (c) total stress circle at soil–platen interface.

circle in Figure 8.9(c), the α stress characteristics meet the top platen tangentially and β characteristics meet it at right angles. The opposite occurs along the lower interface.

4. Stress singularities occur at points A and C, which are therefore centres from which stress characteristics radiate.

5. The first step in the construction is to draw radiating straight line characteristics from points A and C at suitable angular intervals (in this case 15° has been chosen), and link these with the circular arc characteristics FG, FH which have centres at A and C respectively.

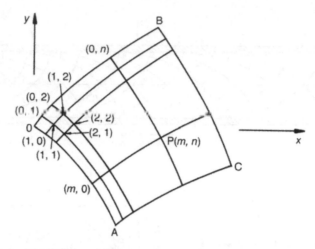

Figure 8.10 Progressive construction of stress characteristics.

6. The family of characteristics can now be constructed progressively in the *x* direction, ensuring that all characteristics cross orthogonally and requirements 1–3 above are satisfied. Various procedures for doing this are described by Hill (1950), including a step-by-step approach based on the diagram in Figure 8.10.

In Figure 8.10 the point (1, 1) is determined from the two points (0, 1) on the known characteristic 0B and (1, 0) on the known characteristic 0A. The initial slopes $(\lambda_{0,1})$ and $(\lambda_{1,0})$ of the small arcs connecting these points to the point (1, 1) are orthogonal to the known stress characteristics and the final slopes depend on the angular interval chosen. A good approximation consists in replacing each arc by a chord with a slope equal to the mean of the initial and final slopes of the arc. It is then possible to write:

1. for arc (0, 1) to (1, 1),

$$y_{1,1} - y_{0,1} = [\tan\tfrac{1}{2}(\lambda_{1,1} + \lambda_{0,1})](x_{1,1} - x_{0,1}) \qquad (8.18)$$

2. for arc (1, 0) to (1, 1),

$$y_{1,1} - y_{1,0} = -[\cot\tfrac{1}{2}(\lambda_{1,1} + \lambda_{1,0})](x_{1,1} - x_{1,0}) \qquad (8.19)$$

$x_{1,1}$ and $y_{1,1}$ can be determined by solving these equations or, with reasonable accuracy, by geometrical construction. As the net moves progressively away from the free face ABC in Figure 8.9(a) the stress characteristics achieve a cycloidal shape, for which the length in the horizontal *x* direction is $\pi D/2$.

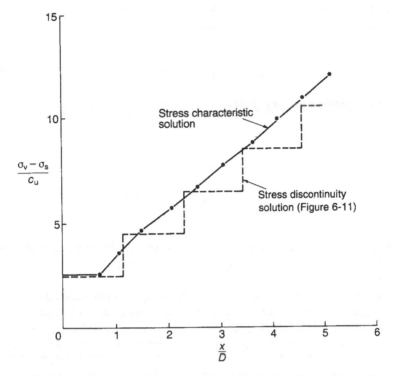

Figure 8.11 Vertical stress distribution between platens and soil for undrained flow of soil between rough platens.

The distribution of direct vertical stress σ_v between the platens and the soil is plotted non-dimensionally in Figure 8.11 as stress $(\sigma_v - \sigma_s)/c_u$ vs x/D. As the length AG is a straight line characteristic, the stress along this length, equal to $D/2^{1/2}$, is constant, and given by

$$s_{A \to G} = s_{A \to C} + 2c_u \frac{\pi}{4}$$

From equation 8.17:

$$\therefore \quad s_{A \to G} = \sigma_s + c_u \left(1 + \frac{\pi}{2}\right) \tag{8.20}$$

From Figure 8.9(c):

$$\sigma_{v,A \to G} = s_{A \to G} = \sigma_s + c_u \left(1 + \frac{\pi}{2}\right) \tag{8.21}$$

For all stress characteristics GP and beyond, the change in slope along the length of the characteristic is $\pi/2$ and thus, from equation 8.3, the change in stress is given by

$$s_P - s_H = s_N - s_G = \pi c_u \tag{8.22}$$

Thus, from equations 8.21 and 8.22:

$$\sigma_{v,N} = s_N = s_{A \to G} + \pi c_u$$

$$\sigma_{v,N} = s_N = s_{A \to G} + \pi c_u$$

$$\therefore \quad \sigma_{v,N} = \sigma_s + c_u \left(1 + \frac{3\pi}{2} \right) \tag{8.23}$$

As the stress characteristics approach cycloidal shape with increasing distance x away from the free face ABC, the pressure distribution in Figure 8.11 tends to a slope of 2.0.

The pressure distribution for the stress discontinuity solution obtained in Section 6.6, and depicted in Figure 6.11, is also reproduced in Figure 8.11. It can be seen that the pressure distribution given by the characteristic solution lies above the average line through the stepped pressure line given by the discontinuities solution.

The above solution assumes that the soil undergoing plastic flow extends indefinitely in the x direction, whereas in any practical case the lateral extent will be limited. In this case an anomaly arises in mid-Section XYZ (half Section) as shown in Figure 8.12, and it must be assumed that this mid-Section acts rigidly, but loses soil to the plastic region as compression progresses. For further discussion on this the reader is referred to specialist texts on plasticity. A solution presented by Chakrabarty (1987) is shown in Figure 8.12. In common with all solutions derived for plasticity of metals, Chakrabarty's solution is based on a network of slip lines, but in fact for metals, these are coincident with stress characteristics (Section 8.5) and the solution shown in Figure 8.12 can also be considered a valid solution for undrained plastic flow in soils based on strength characteristics.

8.3 Drained stress characteristics

The two-dimensional effective stresses at a point in a mass of soil experiencing shear failure under drained conditions are shown in Figure 8.13(a) and may be represented by the Mohr circle in Figure 8.13(b). As the soil is at failure the stress circle touches the c', ϕ' envelopes. The pole point for planes P is easily found by a line through σ'_1 in Figure 8.13(b) parallel to the direction of the plane on which σ'_1 acts in Figure 8.13(a). The two lines through P which pass through the intersections of the stress circle and the strength envelopes give

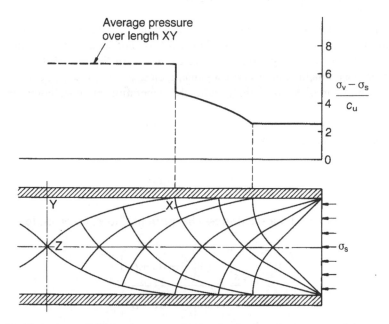

Figure 8.12 Chakrabarty (1987) solution at mid-Section for flow between parallel platens.

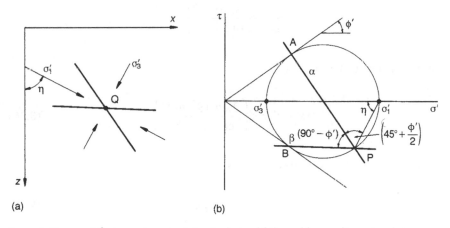

Figure 8.13 α and β stress characteristics in drained failure: (a) two-dimensional stress system; (b) effective stress circle.

the directions of the α (positive shear) and β (negative shear) characteristics. It can be seen from Figure 8.13(b) that the α, β, characteristics intersect at angles of $(\frac{1}{2}\pi \pm \phi')$.

The simplest sets of characteristics which intersect at angles of $(\frac{1}{2}\pi \pm \phi')$ are parallel straight lines as shown in Figure 8.14(a) and logarithmic spirals crossing straight lines radiating from a centre as shown in Figure 8.14(b). A number of simple problems can be addressed using these two basic sets of characteristics, but more complex problems require the generation of more complex sets of characteristics.

Where a small change in stress direction $d\theta$ occurs across a discontinuity, equation 7.15 can be written

$$d\theta = \frac{\pi}{2} - d\Lambda \tag{8.24}$$

and equation 7.17, for a small stress change from s'_0 to s', becomes, for $c' = 0$:

$$\frac{s'}{s'_0} = \frac{\sin\left(\frac{\pi}{2} - d\theta + \delta'\right)}{\sin\left(\frac{\pi}{2} - d\theta - \delta'\right)}$$

or

$$\frac{s'}{s'_0} = \frac{\cos(d\theta - \delta')}{\cos(d\theta + \delta')} \tag{8.25}$$

from which it can be shown that

$$s' - s'_0 = s'_0\left(\frac{2\sin d\theta \cdot \sin \delta'}{\cos(d\theta + \delta')}\right) \tag{8.26}$$

As $d\theta \to 0$, $\sin d\theta \to d\theta$, $\cos d\theta \to 1$, $\delta' \to \phi'$,

$$\therefore \frac{ds'}{s'} = 2d\theta \tan \phi' \tag{8.27}$$

Integrating along a finite length of characteristic:

$$s' - s'_0 \exp[2(\theta - \theta_0) \tan \phi'] \tag{8.28}$$

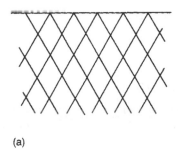

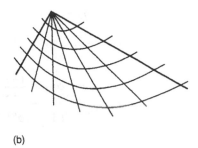

(a)   (b)

Figure 8.14 Drained stress characteristics (a) parallel straight lines; (b) radiating straight lines and logarithmic spirals.

It follows from equation 8.28 that stresses remain constant along a straight characteristic.

Equation 8.28 can also be written

$$s'_C = s'_B \exp(2\theta_D \tan \phi') \tag{8.29}$$

where θ_D is the angle subtended by a zone D of radiating straight stress characteristics and log spirals separating zones of constant stress B and C.

8.3.1 Smooth strip footing

Consider a uniformly loaded strip footing, as shown in Figure 8.15(a), at the point of failure with an applied vertical pressure of q_F. A uniform surcharge loading p_S acts on the surface adjacent to the footing. The soil has the drained strength parameters c', ϕ'.

From Figure 8.15(b):

$$s'_B = \frac{c' \cos \phi' + p_S}{1 - \sin \phi'} \tag{8.30}$$

$\theta_D = \pi/2$, so that equation 8.29, with the inclusion of the c' term, becomes:

$$s'_C + c' \cot \phi' = (s'_B + c' \cot \phi') \exp(\pi \tan \phi') \tag{8.31}$$

i.e.

$$(s'_C + c' \cot \phi') = \left(\frac{c' \cos \phi' + p_S}{1 - \sin \phi'} \right) \exp(\pi \tan \phi') \tag{8.32}$$

but, from Figure 8.15(b):

$$q_F = (s'_C + c'\cot\phi')(1 + \sin\phi') - c'\cot\phi' \qquad (8.33)$$

Combining equations 8.32 and 8.33 gives

$$q_F = c'\cot\phi'\left[\left(\frac{1+\sin\phi}{1-\sin\phi}\right)\exp(\pi\tan\phi') - 1\right]$$
$$+ p_s\left(\frac{1+\sin\phi'}{1-\sin\phi'}\right)\exp(\pi\tan\phi') \qquad (8.34)$$

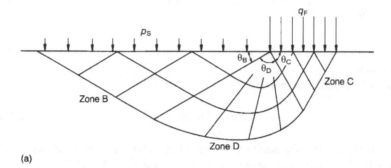

(a)

(b)

Figure 8.15 Smooth strip footing under vertical loading: (a) drained stress characteristics; (b) effective stress circles.

The conventional bearing capacity expression for a c', ϕ' soil is

$$q_F = c'N_c + \gamma D_e N_q + \tfrac{1}{2}\gamma B N_\gamma \qquad (8.35)$$

In the above theory the soil is assumed to be weightless, and therefore $\tfrac{1}{2}\gamma B N_\gamma = 0$. In fact this only applies below foundation level, as a surface loading is assumed. The weight of soil above foundation level becomes $\gamma D_e = p_S$, and thus from equation 8.34:

$$N_c = \cot\phi'\left[\left(\frac{1+\sin\phi}{1-\sin\phi}\right)\exp(\pi\tan\phi')-1\right] \qquad (8.36)$$

$$N_q = \left(\frac{1+\sin\phi'}{1-\sin\phi'}\right)\exp(\pi\tan\phi') \qquad (8.37)$$

Thus

$$N_c = (N_q - 1)\cot\phi' \qquad (8.38)$$

Values of N_c, N_q for $\phi' = 20°$, $30°$ and $40°$ are given in Table 8.2.

As discussed in Section 7.6.3, no closed form solution exists for evaluating the bearing capacity factor N_γ in equation 8.35. An approximate method of evaluating N_γ is given in Section 7.6.3, using stress discontinuities and balancing active and passive forces across a vertical face below an edge of the loaded area. An alternative approximate method, which can use the stress characteristic solution, is presented by Bolton (1979), who assumed the soil below foundation level to a depth of $B/2$ to act as a surcharge. Thus

$$q_F(2)+\frac{\gamma B}{2} = \frac{\gamma B}{2}\left(\frac{1+\sin\phi'}{1-\sin\phi'}\right)\exp(\pi\tan\phi') \qquad (8.39)$$

i.e.

$$q_F(2) = \frac{\gamma B}{2}(N_q - 1) \qquad (8.40)$$

or

$$N_\gamma = N_q - 1 \qquad (8.41)$$

Values of N_γ from equation 8.41 are given in Table 8.2, and it can be seen, by referring to Table 7.3, that the agreement with the Sokolovski values is slightly poorer than given by the method used in Section 7.6.3.

Table 8.2 Values of N_c, N_q and N_γ

ϕ'	N_c Eq. 8.36	N_q Eq. 8.37	N_γ	
			Eq. 8.41	Sokolovski
20°	15	6.4	5.4	3.2
30°	30	18	17	15
40°	76	64	63	86

Table 8.3 Values of q_{IF}/p_S and θ_D

ϕ'	$\omega = 0$		$\omega = 10°$		
	q_{IF}/p_S	θ_D	q_{IF}/p_S	Ω	θ_D
20°	6.4	90°	4.7	30.5°	69.8°
30°	18	90°	13	20.3°	74.8°
40°	64	90°	43	15.7°	77.2°

8.3.2 Strip footing with inclined loading on weightless soil

Consider a strip footing, as shown in Figure 8.16(a), with uniform loading at failure q_{IF} inclined at angle ω to the vertical. A uniform surcharge loading p_S acts on the surface adjacent to the footing. The soil has drained strength parameters $c' = 0$, ϕ'.

The applied stress q_{IF} plots as a point Q on a line through the origin inclined at ω to the $\tau = 0$ axis on a Mohr stress diagram, as shown in Figure 8.16(b). A circle can then be completed through point Q, touching the strength envelopes, representing the stresses in zone C. The stresses in zone B are represented by a circle touching the two envelopes and passing through the minor principal stress p_S on the $\tau = 0$ axis. It can be seen that the α, β stress characteristics in zone B are straight lines inclined at angles of $\frac{1}{2}(90° - \phi')$ to the horizontal. From the geometry of Figure 8.16(b):

$$\theta_B = \tfrac{1}{2}(90° - \phi')$$
$$\theta_C = \tfrac{1}{2}(90° + \phi' + \Omega + \omega)$$
$$\theta_D = 180° - \theta_B - \theta_C$$
$$\therefore \quad \theta_D = \tfrac{1}{2}(180° - \Omega - \omega).$$

Equation 7.19: $\sin\Omega = \dfrac{\sin\omega}{\sin\phi'}$

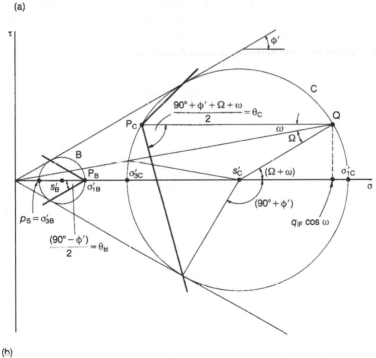

(a)

(b)

Figure 8.16 Strip footing under inclined loading: (a) drained stress characteristics; (b) effective stress circles.

The value of q to cause failure can be found from Figure 8.16(b):

$$q_{IF} \cos\omega = s'_C[1 + \sin\phi' \cos(\Omega + \omega)] \qquad (8.42)$$

where

$$s'_C = s'_D \exp(2\theta_D \tan\phi')$$

and

$$s'_B = \frac{p_S}{1 - \sin\phi'}$$

Table 8.3 gives values of q_{IF}/p_S for $\omega = 0$ and $10°$, and $\phi' = 20°, 30°, 40°$.

8.3.3 Silo problem

(a) Smooth walls

In Section 7.7.1 the wall pressure on a silo with a lower Section inclined at 45° was calculated using both a single discontinuity and two discontinuities. It was shown that for smooth walls the change in direction of the principal stresses, and hence θ_D, was 45°. Thus, from equation 8.29:

$$s'_C = s'_B \exp\left(\frac{\pi}{2}\tan\phi'\right)$$

putting

$$s'_B = \frac{\sigma'_{nW}(B)}{1 - \sin\phi'}$$

and

$$\sigma'_{nW}(C) = s'_C(1 + \sin\phi')$$

which gives $\sigma'_{nW}(C)/\sigma'_{nW}(B) = 7.4$ for $\phi' = 30°$ and 17.2 for $\phi' = 40°$. These values are shown, together with the solutions for a single discontinuity and two discontinuities, in Table 7.4.

(b) Rough walls

Assuming walls with a friction angle of $\phi'/2$ and the lower Section having a slope of 45°, it was shown in Section 7.7.2 that the change in major stress direction is 13.8° (0.241 rad), which is also the angle of the fan of characteristics. Thus

$$s'_C - s'_B \exp(2 \times 0.241 \times \tan 30°)$$
$$= 1.321$$

This differs very little from the values of 1.316 and 1.319 given by, respectively, a single discontinuity and two discontinuities (Section 7.7.2).

EXAMPLE 8.3 STRESS CONCENTRATION IN SILO WALL

Solve Example 7.3 using stress characteristics. Draw the stress characteristics.

Solution

As shown in the solution for Example 7.3, the major principal stress changes direction by 25.4° (0.443 rad) from zone B to zone C.

$$\text{Equation 8.29: } s'_C = s'_B \exp(2 \times 0.443 \tan 32°)$$
$$\therefore \quad s'_C = 1.74 s'_B$$

$$\text{Figure 7.11(b): } s'_B = \sigma'_{nw}(B) + s'_B \sin\phi' \cos(\Omega - \omega')$$
$$= \sigma'_{nw}(B) + s'_B \sin 32° \cos 26°$$

Thus:

$$s'_B = 1.91 \sigma'_{nw}(B)$$

$$\text{Figure 7.11(c): } \sigma'_{nw}(C) = s'_C + s'_C \sin\phi' \cos(\Omega + \omega')$$
$$= s'_C(1 + \sin 32° \cos 23.2°)$$
$$= 1.487 s'_C$$

$$\therefore \quad \sigma'_{nw}(C) = \sigma'_{nw}(B)[1.74 \times 1.91 \times 1.49]$$
$$\therefore \quad \sigma'_{nw}(C) = 4.95 \sigma'_{nw}(B)$$

This compares with a value of 4.91 obtaining in the solution of Example 7.3 using two assumed stress discontinuities. The stress characteristics and stress diagrams are shown in Figure 8.17.

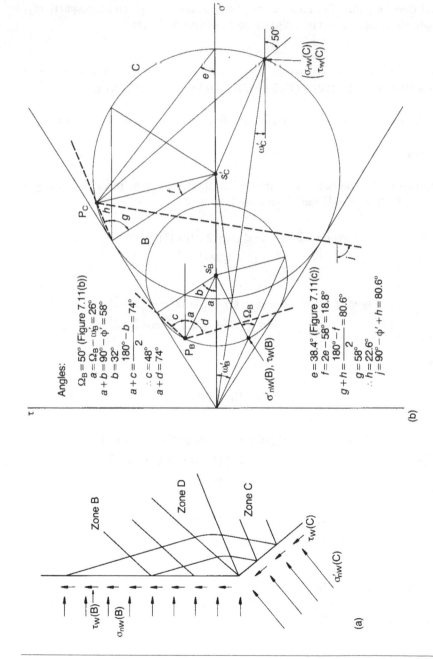

Angles:

$\Omega_B = 50°$ (Figure 7.11(b))
$a = \Omega_B - \omega_B' = 26°$
$a + b = 90° - \phi' = 58°$
$b = 32°$
$a + c = \dfrac{180° - b}{2} = 74°$
$\therefore c = 48°$
$a + d = 74°$

$e = 38.4°$ (Figure 7.11(c))
$f = 2e - 58° = 18.8°$
$g + h = \dfrac{180° - f}{2} = 80.6°$
$g = 58°$
$\therefore h = 22.6°$
$j = 90° - \phi' + h = 80.6°$

(b)

Figure 8.17 Example 8.3: (a) stress characteristics; (b) stress circles.

8.4 Rankine limiting stress states

Rankine (1857) considered the stability of earth masses, assuming them to have frictional strength only. He defined the friction angle ϕ as the *angle of repose* and the coefficient of friction as the tangent of this angle. For an earth mass with a horizontal surface he took the vertical stress σ_v at a depth z to be a principal stress given by:

$$\sigma_v = \gamma z$$

where γ is the unit weight of the soil. He then showed that the condition of maximum stress obliquity restricted the horizontal principal stress σ_h to values between upper and lower limits given by the expression

$$\sigma_h = \gamma z \left(\frac{1 \pm \sin \phi}{1 \mp \sin \phi} \right) \tag{8.43}$$

which correspond to the classical lower bound plasticity solutions.

It follows from equation 8.43 that the planes on which the maximum stress obliquity acts (seen as straight lines in Section for plane strain) are at an angle of $(45° - \phi/2)$ to the vertical when σ_h has its minimum value, and $(45° + \phi/2)$ to the vertical when σ_h has its maximum value. These are the active and passive cases respectively.

As the soil is not weightless, these straight lines cannot be stress characteristics. However, dividing the stresses by γz transforms them into a dimensionless form, and each specific stress condition is represented by one dimensionless stress circle throughout the mass. The lines along which the maximum stress obliquity act are thus also non-dimensional stress characteristics.

8.5 Slip lines

In the previous Sections of this chapter, solutions have been obtained for plastic deformation problems simply by satisfying stress equilibrium and non-violation of the failure criterion within the deforming mass. These give a lower bound on the forces deforming the mass, and are thus conservative; but at the same time they can provide excellent insights into behaviour of the deforming mass. They provide no information on the actual deformations incurred or on the work done in deforming the mass.

Consideration of the deformations should in general lead to better solutions. There is still considerable controversy over the derivation of the displacement equations, and it is beyond the scope of this book to derive such equations, or to discuss them in any detail (see e.g. Houlsby and Wroth, 1982, for such discussion). Basically it is envisaged that sliding occurs on surfaces which, in plane strain viewed in Section, become a pattern of slip lines with a configuration similar to,

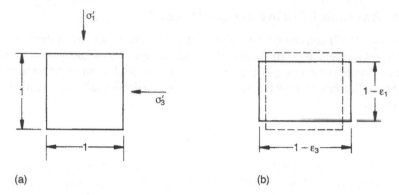

Figure 8.18 Plastic deformation: (a) principal stress on an element; (b) corresponding coaxial strains.

but not necessarily identical with, the stress characteristics. In order to fix the orientation of the pattern, coaxiality is commonly assumed; that is, as shown in Figures 8.18(a) and (b), the axes of major principal stress and major principal strain are assumed to coincide. This approach was adopted by Davis (1968).

A further assumption is required to complete the pattern, and an approach which has perhaps attracted most support is to identify slip lines as lines along which zero direct strain occurs, that is lines of 'zero extension'. Roscoe (1970) proposed the assumption of slip lines as zero extension lines on the basis of experimental evidence from model studies of sand deformations behind retaining walls subjected to small rotations (James *et al.*, 1972). X-ray observations of lead shot grids embedded in the sand allowed precise measurements to be made of deformations within the sand. Figure 8.19(a) shows stress characteristic directions for the deforming element in Figure 8.18, and the corresponding slip line directions

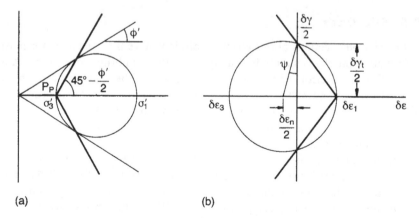

Figure 8.19 Plastic deformation: (a) effective stress circle showing directions of stress characteristics; (b) strain circle showing direction of slip lines.

for small strain increments $\delta\varepsilon_1$ and $\delta\varepsilon_2$ are shown in Figure 8.19(b). It can be seen that

$$\tan\psi = -\frac{\delta\varepsilon_n}{\delta\gamma_t} \tag{8.44}$$

where ψ is the dilation angle (Section 1.6), $\delta\varepsilon_n$ and $\delta\gamma_t$ are the increments of normal strain and shear strain respectively along the slip lines. Note that

$$\delta\varepsilon_n = \delta\varepsilon_V = \delta\varepsilon_1 + \delta\varepsilon_3$$

as direct strains are zero in the directions of the slip lines. An illustrative relationship between stress characteristics and slip lines at a point Q in a plastically deforming soil is depicted in Figure 8.20, and typical fields of stress characteristics and slip lines in sand behind a wall rotating about its top into the sand are shown in Figure 8.21 (James *et al.*, 1972).

In the particular case where $\psi = \phi'$ the meshes of stress characteristics and slip lines become coincident. That is, the ratio of normal strain to shear strain along the slip line is equal to the ratio of normal stress to shear stress. This is known as the 'associated' flow rule. In most cases, in sand, the associated flow rule will not hold.

It follows from the above that in making stability analyses based on assumed slip lines the stress ratio to apply is not the maximum stress ratio (except where the associated flow rule holds), but the actual stress ratio on the slip surface. This can be deduced from the stress characteristics, but in some problems such as the active case for a retaining wall, the slip line field may extend outside the stress characteristic field and consequently the stresses on part of the slip line field are not known. According to Roscoe (1970), the stress ratio on a slip line can be measured directly in a direct shear test such as the simple shear apparatus in which the zero extension lines are horizontal.

8.6 Undrained deformation

The application of slip line theory can be simply illustrated by the case of a foundation with a smooth base causing failure in a saturated clay, assuming undrained behaviour, for which $\varepsilon_V = 0$ and hence $\psi = 0$. The strain circle for a soil deforming in plane strain in this way is shown in Figure 8.22, and it can be seen that the lines of zero extension are orthogonal and inclined at 45° to the major principal strain increment ε_1. As coaxiality is assumed, they are also inclined at 45° to the σ_1 direction, and thus coincide with the directions of the stress characteristics.

While this coincidence of slip lines and stress characteristics for undrained deformation is a useful assumption for proceeding to solutions of plasticity problems, it should be noted that it conflicts with the observations in triaxial tests,

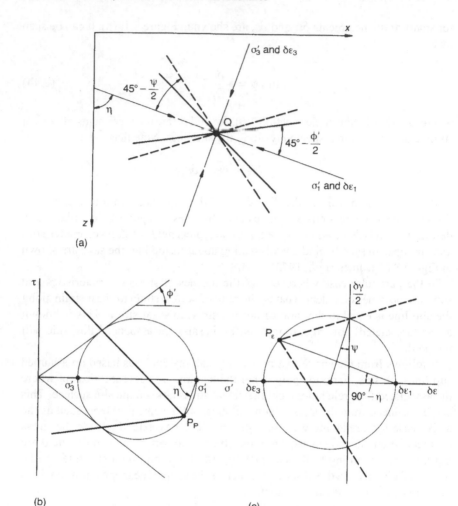

Figure 8.20 Plastic deformation: (a) illustrative relationship between stress
characteristics and slip lines; (b) corresponding effective stress circle;
(c) corresponding strain circle.

discussed in Section 2.6, that rupture planes are inclined at $(45° - \phi'/2)$ to the
direction of the major principal stress.

A kinematically possible set of slip lines for a soil undergoing undrained plastic
deformation below a smooth foundation is shown in Figure 8.23(a), in which:

1. block cdl is forced vertically downwards as a rigid body by an amount u;
2. blocks abg and efq are forced upwards and outwards as rigid bodies sliding
    on planes ag and fq respectively;
3. uniform deformations occur in the two fan zones.

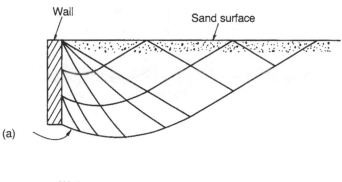

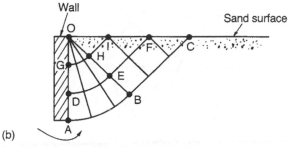

Figure 8.21 Retaining wall rotating about its top into sand: (a) stress characteristics; (b) slip lines. (After James *et al.*, 1972.)

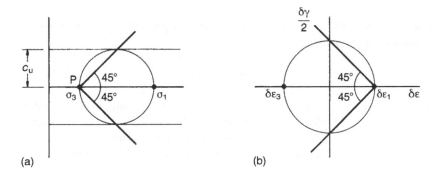

Figure 8.22 Undrained deformation: (a) total stress circle showing directions of stress characteristics; (b) strain circle showing directions of slip lines.

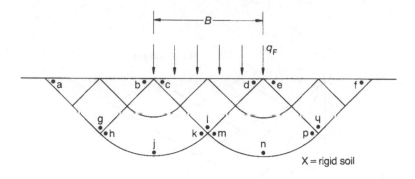

(a)

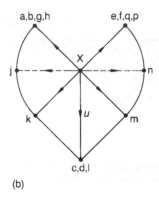

(b)

Figure 8.23 Vertically loaded strip footing with smooth base: (a) slip lines; (b) displacement diagram.

The external work W_e done by the foundation pressure q_F for unit length of foundation is given by

$$W_e = q_F B u \qquad (8.45)$$

In order to obtain a solution, equation 8.45 must be equated to the internal work W_i, which can be obtained by consideration of the displacement diagram shown in Figure 8.23(b). A particular point to note in this diagram is the relative movements of points klm, which are dictated by the fact that the slip lines shown are lines of zero extension.

As blocks abg, cdl and efq are rigid, no work is done within the blocks; but work W_{il} is expended by sliding along planes ag and fq and by sliding along the slip lines separating block cdl from the two fan zones. The displacements along each of these planes is $u/2^{1/2}$ and the force acting along the planes is c_u by the length, $B/2^{1/2}$ of each of the planes. Thus

$$W_{il} = 4c_u \frac{B}{2^{1/2}} \frac{u}{2^{1/2}}$$

$$= 2Bc_u u \tag{8.46}$$

As the displacements normal to the linear slip lines bounding each fan zone are uniform as shown in Figure 8.24(a), and not proportional to the distance from the apex, the fan zones are not simply rotating about their apices. Distortions must occur within the fan zone and these are assumed to be uniform. The work expended by these distortions can be calculated by considering an infinitesimally small sector of fan zone, as shown in Figure 8.24(b).

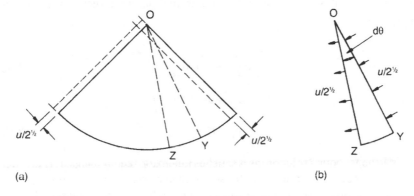

(a) (b)

Figure 8.24 Fan zone: (a) boundary movements; (b) distortion of an infinitely small segment.

$$\text{Displacement along OY} = \frac{u}{2^{1/2}} d\theta$$

$$\therefore \text{ Work done along OY} = c_u \frac{B}{2^{1/2}} \frac{u}{2^{1/2}} u$$

Integrating over the fan,

$$\text{Work done} = \frac{\pi}{4} Bc_u u$$

$$\text{Displacement along YZ} = \frac{u}{2^{1/2}}$$

$$\text{Work done along YZ} = c_u \frac{B}{2^{1/2}} \frac{u}{2^{1/2}} u$$

Integrating over the fan,

$$\text{Work done} = \frac{\pi}{4} Bc_u u$$

Thus the total work W_{i2} expended by distortions within the two fans and slipping along the curved boundaries is

$$W_{i2} = 2\left(\frac{\pi}{4} Bc_u u + \frac{\pi}{4} Bc_u u\right) = \pi Bc_u u \tag{8.47}$$

Equating external work (equation 8.45) with internal work (equations 8.46 and 8.47) gives

$$W_e = W_{i1} + W_{i2} \tag{8.48}$$

$$\therefore q_F Bu = Bc_u (2 + \pi)u$$

$$\therefore q_F = (2 + \pi)c_u \tag{8.49}$$

This is an upper bound solution, as it satisfies kinematic requirements. It is identical to the lower bound solution given by stress characteristics in equation 8.11. It is therefore the correct solution for a soil assumed to be deforming plastically.

Appendix: Symbols

Superscript

′ (prime)   Effective stress parameter

Superscript or subscript

u   Undrained total stress parameter

Subscripts

1, 2, 3   Respectively major, intermediate and minor principal stress or strain
directions
a   Axial direction
f   Failure value
h   Horizontal direction
r   Radial direction
v   Vertical direction
x, y, z x, y, z directions

Roman symbols

A (i) Area; (ii) pore pressure coefficient
B (i) Width of foundation or surface loading; (ii) constant in Johnston
expression for strength; (iii) pore pressure coefficient
C Triaxial compression
CD   Consolidated drained triaxial test
CU   Consolidated undrained triaxial test
D_e Embedded depth of a foundation
E (i) Young's modulus; (ii) triaxial extension
E_0, E_{max} Small strain Young's modulus
E_h, E_r Horizontal/radial Young's modulus in cross-anisotropic soil or rock
E_v, E_a Vertical/axial Young's modulus in cross-anisotropic soil or rock

F	(i) Force; (ii) factor of safety
F_A	Active force
F_{hA}	Horizontal active force
F_{hP}	Horizontal passive force
F_{hW}	Horizontal force on unit length of retaining wall
F_P	Passive force
G	Shear modulus
G_0, G_{max}	Small strain shear modulus
G_{hh}, G_{rr}	Shear modulus in the horizontal/radial plane in cross-anisotropic soil or rock
G_{hv}, G_{ra}	Shear modulus in the vertical/axial plane in cross-anisotropic soil or rock
G_{vh}, G_{ar}	Shear modulus in the vertical/axial plane in cross-anisotropic soil or rock
H	Height of retaining wall
I_D	Relative density of sand
I_P	Plasticity index
I_R	Bolton dilatancy index
JRC	Compressive strength of rock joint wall
K_0	Coefficient of lateral earth pressure
K_A	Coefficient of active earth pressure
K_f	Ratio of t/s' at failure
K_{nc}	K_0 for normally consolidated soil
K_P	Coefficient of passive earth pressure
M	(i) Critical state strength parameter $(= q_{cs}/p'_{cs})$; (ii) constant in Johnston expression for strength
N	(i) Force normal to a plane; (ii) number of fractal chords
N_c	Bearing capacity factor
N_q	Bearing capacity factor
N_γ	Bearing capacity factor
OCR	Overconsolidation ratio
P	Pole point
R	Schmidt hammer rebound on dry unweathered rock
S	Constant in Hoek and Brown strength expression for rock
T	Shear force on a plane
T_0	Uniaxial tensile strength of uncracked brittle material
UU	Undrained triaxial test
V_S	Shear wave velocity
c	Cohesion intercept
c'_e	Hvorslev effective stress cohesion
c_j	Cohesion intercept for rock joint or discontinuity
c_r	Cohesion intercept for intact rock
c'_r	(i) Effective stress c_r for intact rock; (ii) effective stress residual cohesion intercept for soil

c'_R	Gibson modified cohesion intercept
c_u	Undrained shear strength
c_{uh}	c_u for specimen sampled horizontally
c_{uv}	c_u for specimen sampled vertically
c_W	Adhesion between soil and wall
e	Voids ratio
h	Height of ground surface above specified point
h_w	Height of water table or phreatic surface above specified point
i	Dilation angle for rock joint or discontinuity
k'	Cohesion intercept given by K_f line $= c' \cos \phi'$
l	Length
m	Constant in Hoek and Brown strength expression for rock
n	Number of stress discontinuities
n'	Effective stress stiffness ratio $\dfrac{E_r' \text{ or } E_h'}{E_a' \text{ or } E_v'}$
n^u	Undrained total stress stiffness ratio $\dfrac{E_r^u \text{ or } E_h^u}{E_a^u \text{ or } E_v^u}$
p	Mean stress ($= \frac{1}{3}(\sigma_1 + \sigma_2 + \sigma_3)$; $= \frac{1}{3}(\sigma_a + 2\sigma_r)$ in triaxial test)
p'_{cs}	Critical state value of p'
p_S	Surface surcharge pressure on soil adjacent to a foundation
q	(i) Deviator stress ($= (\sigma_1 - \sigma_3) = (\sigma_1' - \sigma_3')$); (ii) deviator stress in triaxial test ($= (\sigma_a - \sigma_r) = (\sigma_a' - \sigma_r')$)
q_{cs}	Critical state value of q
q_F	Foundation or surface loading pressure causing failure
q_S	Surcharge pressure on soil surface behind retaining wall
r	(i) Radius of circle; (ii) Schmidt hammer rebound on weathered saturated rock in joint wall
s	Stress parameter $= \frac{1}{2}(\sigma_1 + \sigma_3) = \frac{1}{2}(\sigma_a + \sigma_r)$ in triaxial test
s_0	Initial s on total stress path
s'_0	Initial s' on effective stress path
s'_e	Value of s' in sample after extrusion: $s'_e = -u'_e$
s'_i	Value of s' after applying cell pressure in triaxial test
t	(i) Stress parameter ($= \frac{1}{2}(\sigma_1 - \sigma_3) = \frac{1}{2}(\sigma_1' - \sigma_3')$); (ii) stress parameter in triaxial test ($= \frac{1}{2}(\sigma_a - \sigma_r) = \frac{1}{2}(\sigma_a' - \sigma_r')$)
u	(i) Pore pressure; (ii) small displacement
u_0	Pore pressure at specified point in the ground
u_e	Pore pressure in sample after extrusion
u_i	Pore pressure in sample after application of cell pressure to sealed sample in triaxial cell
w	Water content
w_L	Liquid limit
w_P	Plastic limit
z	Depth below ground level
z_w	Depth of water table below ground level

Greek symbols

Δ	(i) Increment of a quantity; (ii) angle
Ω	(i) Angle; (ii) variable parameter
α	(i) Angle; (ii) stress characteristic along which shear stresses are positive
α'	Slope of K_f line ($= \tan^{-1} \sin \phi'$)
β	(i) Angle between joint or physical discontinuity and plane on which σ_1 acts; (ii) slope of ground surface; (iii) angle to the horizontal of a stress discontinuity; (iv) stress characteristic along which shear stress is negative; (v) an angle
γ	(i) Engineers' shear strain; (ii) unit weight
γ'	Buoyant unit weight
γ_{max}	Maximum shear strain
γ_W	Unit weight of water
δ	Small increment of a quantity
δ'	Effective stress friction angle developed on a stress discontinuity
ε	Direct strain
ε_V	Volumetric strain
ε_{xz}	Pure shear strain in xz plane
ε_{zx}	Pure shear strain in xz plane $= \varepsilon_{xz}$
η	Angle
θ	Angle
λ	Angle associated with stress characteristics
ν	Poisson's ratio
ν_0	Small strain Poisson's ratio
ν_{hh}, ν_{rr}	Ratio of strain in horizontal/radial direction to an imposed orthogonal horizontal/radial strain for cross-anisotropic soil or rock
ν_{hv}, ν_{ra}	Ratio of strain in vertical/axial direction to an imposed horizontal/radial strain for cross-anisotropic soil or rock
ν_{vh}, ν_{ar}	Ratio of strain in horizontal/radial direction to an imposed vertical/axial strain for cross-anisotropic soil or rock
ρ	Density
σ	Direct total stress
σ'	Direct effective stress
σ'_{1N}	σ'_1 / σ_c for rock
σ'_{3N}	σ'_3 / σ_c for rock
σ_c	Unconfined compressive strength of intact rock
σ'_c	Triaxial consolidation pressure
σ_{cm}	Unconfined compressive strength of rock mass
σ_{cp}	Triaxial cell pressure
σ_{h0}	Horizontal direct total stress at a specified point in the ground
σ'_{h0}	Horizontal direct effective stress at a specified point in the ground
σ_{hW}	Horizontal total stress on vertical retaining wall
σ'_{hW}	Horizontal effective stress on vertical retaining wall

σ_n	Direct total stress normal to a plane
σ'_n	Direct effective stress normal to a plane
$\sigma_{n\theta}$	Normal stress on plane at angle θ to plane on which σ_1 acts
σ_s	Constraining stress on plastic material squeezed between parallel platens
σ_t	Unconfined tensile strength of intact rock
σ_{tm}	Unconfined tensile strength of rock mass
σ_{v0}	Vertical direct total stress at a specified point in the ground
σ'_{v0}	Vertical direct effective stress at a specified point in the ground
τ	Shear stress
τ_f	Shear stress at failure (shear strength)
τ_{fN}	τ_f/σ_c for rock
τ_{jf}	Shear strength of rock joint or discontinuity
τ_{max}	Maximum shear stress
τ_{min}	Minimum shear stress
τ_{rf}	Shear strength of intact rock
τ_{xy}	Shear stress on yz plane acting in y direction
τ_{xz}	Shear stress on yz plane acting in z direction
τ_{yx}	Shear stress on xz plane acting in x direction
τ_{yz}	Shear stress on xz plane acting in z direction
τ_{zx}	Shear stress on xy plane acting in x direction
τ_{zy}	Shear stress on xy plane acting in y direction
τ_θ	Shear stress on plane at angle θ to the plane on which σ_1 acts
ϕ	Angle of friction or shearing resistance
ϕ'	Effective stress angle of friction or shearing resistance
ϕ'_b	Effective stress angle of friction for smooth rock joint or discontinuity
ϕ_{cu}	Consolidated undrained angle of shearing resistance
ϕ'_e	Hvorslev effective stress angle of friction
ϕ'_i	Instantaneous effective stress angle of friction for rock
ϕ_j	Total stress angle of friction for rock joint or discontinuity
ψ_r	Total stress angle of friction for intact rock
ϕ'_R	Gibson modified effective stress angle of shearing resistance
ϕ'_r	(i) Effective stress residual angle of shearing resistance in soils; (ii) effective stress angle of friction in intact rock
ϕ_u	Undrained angle of shearing resistance (usually taken to be zero for saturated clay)
ψ	Dilatancy angle for soils
ω	Inclination to vertical of foundation loading
ω'	Effective stress angle between soil and wall

References

Abbot, M.B. (1966) *An Introduction to the Method of Characteristics*, Thames and Hudson, London.

Aldrich, M.J. (1969) Pore pressure effects on Berea sandstone subject to experimental deformation. *Geol. Soc. Am. Bull.*, **80**, 1577–86.

Atkinson, J.H. (1975) Anisotropic elastic deformations in laboratory tests on undisturbed clay. *Géotechnique*, **25**(2), 357–74.

Atkinson, J.H. (1981) *Foundations and Slopes*, McGraw-Hill, Maidenhead.

Atkinson, J.H. (2000) Non-linear soil stiffness in routine design. *Géotechnique*, **50**(5), 487–507.

Atkinson, J.H. and Richardson, D. (1987) The effect of local drainage in shear zones on the undrained strength of overconsolidated clays. *Géotechnique*, **37**(3), 393–403.

Atkinson, J.H., Richardson, D. and Robinson, P.J. (1987) Compression and extension of K_0 normally consolidated kaolin clay. *ASCE J. Geotech. Engng*, **113**(12), 1468–82.

Balmer, G. (1952) A general analytical solution for Mohr's envelope. *Proc. Am. Soc. Test. Mater.*, **52**, 1260–71.

Barton, N.R. (1973) Review of a new shear strength criterion for rock joints. *Engg Geology*, **7**, 287–332.

Barton, N.R. (1986) Deformation phenomena in jointed rock. *Géotechnique*, **36**(2), 147–67.

Barton, N. and Bandis, N. (1982) Effects of block size on the shear behaviour of jointed rock. *23rd U.S. Symposium on Rock Mechanics*, Berkeley, CA, pp. 739–60.

Barton, N. and Choubey, V. (1977) The shear strength of rock joints in theory and practice. *Rock Mechanics*, **10**, 1–54.

Bishop, A.W. and Garga, V.K. (1969) Drained tension tests on London Clay. *Géotechnique*, **14**(2), 309–12.

Bishop, A.W. and Henkel, D.J. (1962) *The Measurement of Soil Properties in the Triaxial Test*, 2nd edn, Edward Arnold, London.

Bishop, J.F.W. (1953) On the complete solution to problems of deformation of a plastic–rigid material. *J. Mech. Phys. Solids*, **2**, 43–53.

Bolton, M. (1979) *A Guide to Soil Mechanics*, Macmillan Press, London.

Bolton, M.D. (1986) The strength and dilatancy of sands. *Géotechnique*, **36**(1), 65–78.

Brady, B.H.G. and Brown, E.T. (1992) *Rock Mechanics in Underground Mining*, 2nd edn, George Allen and Unwin, London.

Brook, N. (1979) Estimating the triaxial strength of rock. *Int. J. Rock Mech. Mining Sci. Geomech. Abst.*, **16**, 261–64.

Brooker, E.W. and Ireland, H.O. (1965) Earth pressures at rest related to stress history. *Can. Geotech. J.*, **2**, 1–15.

Burland, J.B. (1990) On the strength and compressibility of natural clays. *Géotechnique*, **40**(3), 329–78.

Burland, J.B. and Hancock, R.J.R. (1977) Underground car park at the House of Commons: Geotechnical aspects. *Struct. Engr*, **55**, 87–100.

Chakrabarty, J. (1987) *Theory of Plasticity*, McGraw-Hill, New York.

Chandler, R.J. and Skempton, A.W. (1974) The design of permanent cutting slopes in stiff fissured clays. *Géotechnique*, **24**(4), 457–66.

Clayton, C.R.C. and Heymann, G. (2001) Stiffness of geomaterials at small strains. *Géotechnique*, **51**(3), 245–55.

Clegg, D.P. (1981) Model piles in stiff clay. PhD dissertation, Cambridge University.

Coulomb, C.A. (1776) Essai sur une application des règles des maximis et minimis à quelques problèmes de statique relatifs à l'architecture. *Mém. acad. roy. prés. divers savants*, Vol. 7, Paris.

Culmann, C. (1866) *Die Graphische Statik*, Zurich.

Davis, E.H. (1968) Theories of plasticity and failure of soil masses, in *Soil Mechanics, Selected Topics* (ed. I.K. Lee), Butterworths, London, pp. 341–80.

De Lory, F.A. and Lai, H.W. (1971) Variation in undrained shearing strength by semi-confined tests. *Can. Geotech. J.*, **8**, 538–45.

Desrues, J., Chambon, R., Mockni, N. and Mazerolle, F. (1996) Void ratio evolution inside shear bands in triaxial sand specimens, studied by computed tomography. *Géotechnique*, **46**(3), 529–546.

Donath, F.A. (1972) Effects of cohesion and granularity on deformational behaviour of anisotropic rock, in *Studies in Mineralogy and Precambrian Geology* (eds B.R. Doe and D.K. Smith), *Geol. Soc. Am. Mem.*, **135**, 95–128.

Duncan, J.M. and Seed, H.B. (1966) Strength variations along failure surfaces in clay. *ASCE J. SMFE*, **92**(SM6), 51–104.

Finno, R.J., Harris, WW., Mooney, M.A. and Viggiani, G. (1997) Shear bands in plane strain compression of loose sand inside shear bands in triaxial sand specimens, studied by computed tomography. *Géotechnique*, **47**(1), 149–65.

Fugro (1979) Site investigation, full scale pile tests. Madingley, Cambridge. Unpublished report, Fugro Limited, Department of Energy Contract.

Gibson, R.E. (1953) Experimental determination of the true cohesion and angle of internal friction in clays, in *Proceedings 3rd International Conference on Soil Mechanics and Foundation Engineering*, Vol. 1, pp. 126–30.

Goodman, R.E. (1970) The deformability of joints. *Determination of the Insitu Modulus of Rocks. ASTM Special Technical Publication*, No. 477, 174–196.

Griffith, A.A. (1921) The phenomenon of rupture and flow in solids. *Phil. Trans. R. Soc.*, **228A**, 163–97.

Griffith, A.A. (1924) Theory of rupture, in *Proceedings 1st International Congress on Applied Mathematics*. Delft, pp. 55–63.

Hencky, H. (1923) Über einige statisch bestimmte Fälle des Gleichsewichts in plastischen Körpern. *Z. angew. Math. Mech.*, **3**, 241–51.

Henkel, D.J. (1956) Discussion on: Earth movements affecting LTE railway in deep cutting east of Uxbridge. *Proceedings of the Institution of Civil Engineers*, **5**, Part 2, 320–23.

Henkel, D.J. (1971) The relevance of laboratory measured parameters in field studies, in *Proceedings Roscoe Memorial Symposium*, Foulis, pp. 669–75.

Hill, R. (1950) *The Mathematical Theory of Plasticity*, Clarendon Press, Oxford.

Hoek, E. (1968) Brittle fracture of rock, in *Rock Mechanics in Engineering Practice* (eds G. Stagg and O.C. Zienkiewicz), John Wiley and Sons, New York, pp. 99–124.

Hoek, E. (1983) Twenty-third Rankine Lecture – Strength of jointed rock masses. *Géotechnique*, **33**(3), 185–224.

Hoek, E. and Brown, E.T. (1980) Empirical strength criterion for rock masses. *ASCE J. Geotech. Engg Div.*, **106**(GT9), 1013–36.

Houlsby, G.T. and Wroth, C.P. (1982) Direct solutions of plasticity problems in soils using the method of characteristics, in *Proceedings 4th International Conference on Numerical and Analytical Methods in Geomechanics*, Edmonton, Vol. 3, pp. 1059–71.

Hvorslev, M.J. (1937) Uber die Festigkeitseigenschaften gestoerter bindiger Boden. (On the physical properties of undisturbed cohesive soils.) *Ingeniorvidenskabelige Skrifter*, A., No. 45, Copenhagen. English translation (1969) US Waterways Experiment Station.

ISRM Report (1978) Suggested methods for the quantitative description of discontinuities in rock masses. *International Journal of Rock Mechanics and Mining Sciences*, **15**, 319–368.

Jaeger, C. (1972) *Rock Mechanics and Engineering*, Cambridge University Press.

Jaeger, C. and Cook, N.G.W. (1976) *Fundamentals of Rock Mechanics*, 2nd edn, John Wiley and Sons, New York.

Jaeger, J.C. (1960) Shear fracture of anisotropic rocks. *Geol. Mag.*, **97**, 65–72.

Jáky, J. (1944) A nyugali nyomás tényezöje. (The coefficient of earth pressure at rest.) *Magyar Mérnök és Épitész-Egylet Közlönye* (*J. Union of Hungarian Engineers and Architects*), 355–58.

James, R.G., Smith, I.A.A. and Bransby, P.L. (1972) The predictions of stresses and deformations in a sand mass adjacent to a retaining wall, in *Proceedings 5th European Conference on Soil Mechanics and Foundation Engineering*, Madrid, Vol. 1, pp. 39–46.

Johnston, I.W. (1985) Strength of intact geomechanical materials. *ASCE J. Geotech. Engg*, **111**(GT6), 730–49.

Johnston, I.W. and Chiu, H.K. (1984) Strength of weathered Melbourne mudstone. *ASCE J. Geotech. Engg*, **110**(GT7), 875–98.

Kenney, T.C. and Watson, G.H. (1961) Multi-stage triaxial test for determining c' and ϕ' of saturated soils. *Proc. 5th International Conference on Soil Mechanics and Foundation Engineering*, Vol. 1, pp. 191–95.

Ladanyi, B. and Archambault, G. (1977) Shear strength and deformability of filled indented joints, in *Proceedings International Symposium on Geotechnics of Structurally Complex Formations*, Capri, **1**, 317–26.

Lama, R.L. (1978) Influence of clay filling on shear behaviour of joints. *Proceedings 3rd IAEG Congress*, Madrid, **2**, 27–34.

Leavell, D.A., Peters, J.F. and Townsend, F.C. (1982) Engineering properties of clay shales. *US Army W.E.S. Technical Report 5–71–6*. No. 4.

Lee, Y.H., Carr, J.R., Barr, D.J. and Haas, C.J. (1990) The fractal dimension as a measure of roughness of rock discontinuity profiles. *International Journal of Rock Mechanics and Mining Sciences*, **27**(6), 453–64.

Lo, K.Y. (1965) Stability of slopes in anisotropic soils. *ASCE J. SMFE*, **91**(SM4), pp. 85–106.

McLamore, R. and Gray, K.E. (1967) The mechanical behaviour of anisotropic sedimentary rocks. *J. Engg Ind., Trans. Am. Soc. Mech. Engrs B*, **89**, 62–73.

Mandelbrot, B.B. (1983) *The Fractal Geometry of Nature*. Freeman, San Francisco.

Manson, S.M. (1980) An investigation of the strength and consolidation properties of Speswhite Kaolin. Cambridge University Engineering Department Tripos Pt II Project.

Meyerhof, G.G. (1976) Bearing capacity and settlement of piled foundations. *ASCE J. Geotech. Engg Div.*, **102**, 197–228.

Mohr, O. (1882) Ueber die Darstellung des Spannungszustandes und des Deformationszustandes eines Körperelementes und über die Anwendung derselben in der Festigkeitslehre. *Civilengenieur*, **28**, 113–56. (Also *Technische Mechanik*, 2nd edn, 1914.)

Mühlhaus, H.B. and Vardoulakis, I. (1987) The thickness of shear bands in granular materials. *Géotechnique*, **37**(3), 271–83.

Nadarajah, V. (1973) Stress–strain properties of lightly overconsolidated clays. PhD dissertation, University of Cambridge.

NAVFAC (1971) *Design Manual NAVFAC DM-7*, US Department of the Navy.

Nieto, A.S. (1974) Experimental study of the shear stress–strain behaviour of clay seams in rock masses. PhD dissertation, University of Illinois.

Parry, R.H.G. (1956) Strength and deformation of clay. PhD dissertation, University of London.

Parry, R.H.G. (1968) Field and laboratory behaviour of a lightly overconsolidated clay. *Géotechnique*, **18**(2), 151–71.

Parry, R.H.G. (1970) Overconsolidation in soft clay deposits. *Géotechnique*, **20**(4), 442–45.

Parry, R.H.G. (1971) A study of the influence of intermediate principal stress on ϕ' values using a critical state theory. *4th Asian Reg. Conference on Soil Mechanics and Foundation Engineers*, Bangkok, Vol. 1, pp. 159–66.

Parry, R.H.G. (1976) Engineering properties of clay shales. *US Army W.E.S. Technical Report 5-71-6*, No. 3.

Parry, R.H.G. (1980) A study of pile capacity for the Heather platform. *Ground Engng* **13**(2), 26–31.

Parry, R.H.G. (1988) Short-term slipping of a shallow excavation in Gault clay. *Proc. Inst. Civ. Engrs*, Pt 1, **84**, 337–53.

Parry, R.H.G. and Nadarajah, V. (1973) Multistage triaxial testing of lightly overconsolidated clays. *J. Test. Evaluat.*, **1**(5), 374–81.

Parry, R.H.G. and Nadarajah, V. (1974a) Observations on laboratory prepared lightly overconsolidated kaolin. *Géotechnique*, **24**, 345–57.

Parry, R.H.G. and Nadarajah, V. (1974b) Anisotropy in a natural soft clayey silt. *Engng Geol.* **8**(3), 287–309.

Parry, R.H.G. and Wroth, C.P. (1981) Shear properties of soft clay, in *Soft Clay Engineering* (eds E.W. Brand and R.P. Brenner), Elsevier, Amsterdam, pp. 311–64.

Patton, F.D. (1966) Multiple modes of shear failure in rock. *Proceedings 1st ISRM Congress*, Lisbon, Vol. 1, pp. 509–13.

Phien-Wej, N., Shrestha, U.B. and Rantucci, G. (1990) Effect of infill thickness on shear behaviour of rock joints. *Proceedings International Symposium on Rock Joints*, Loen, Norway, pp. 289–94.

Pratt, H.R., Black, A.D. and Brace, W.F. (1974) Friction and deformation of jointed quartz diorite. *Proceedings 3rd ISRM International Congress*, Denver, **2A**, 306–310.

Rankine, W.J.M. (1857) On the stability of loose earth. *Phil. Trans. R. Soc. Lond.*, **147**, 9.

Reades, D.W. and Green, G.E. (1976) Independent stress control and triaxial extension tests on sand. *Géotechnique*, **26**(4), 551–76.

Roscoe, K.H. (1970) The influence of strains in soil mechanics. *Géotechnique*, **20**(2), 129–70.

Rowe, P.W. (1962) The stress–dilatancy relation static equilibrium of an assembly of particles in contact. *Proc. R. Soc.*, 269A, 500–27.

Saada, A.S. and Bianchini, G.F. (1975) Strength of one-dimensionally consolidated clays. *ASCE J. Geotech. Div.*, **101**(GT11), 1151–64.

Schmertmann, J.H. (1975) Measurement of in situ shear strength, in *Proc. Conference In Situ Measurement of Soil Properties, Raleigh, NC*, Vol. 2, American Society of Civil Engineers, pp. 57–138.

Schmidt, B. (1966) Discussion. *Can. Geotech. J.*, **3**(4), 239–42.

Schofield, A.W. and Wroth. C.P. (1968) *Critical State Soil Mechanics*, McGraw-Hill, London.

Seidel, J.P. and Haberfield, C.M. (1995) Towards an understanding of joint roughness. *Rock Mechanics and Rock Engineering*, **28**(2), 69–92.

Shirley, A.W. and Hampton, L.D. (1978) Shear wave measurements in laboratory sediments. *Journal Acoustical Society of America*, **63**(2), 607–13.

Skempton, A.W. (1948) The effective stresses in saturated clays strained at constant volume, in *Proceedings 7th International Conference on Applied Mechanics*, Vol. 1, p. 378.

Skempton, A.W. (1954) The pore pressure coefficients A and B. *Géotechnique*, **4**(4), 143–47.

Skempton, A.W. (1964) Long-term stability of clay slopes. *Géotechnique*, **14**(2), 77–102.

Skempton, A.W. (1967) The strength along discontinuities in stiff clay. *Proceedings Geotechnical Conference*, Oslo, **2**, 29–46.

Sokolovski, V.V. (1965) *Statics of Granular Media*, Pergamon Press, Oxford.

Stroud, M.A. (1971) The behaviour of sand at low stress levels in the simple shear apparatus. PhD dissertation, Cambridge University.

Swanson, S.R. and Brown, W.S. (1971) An observation of loading path dependence of fracture in rock. *Int. J. Rock Mech. Mining Sci. Geomech. Abst.*, **8**, 277–81.

Terzaghi, K. (1936) The shearing resistance of saturated soils and the angle between the planes of shear. *Proceedings 1st International Conference on Soil Mechanics and Foundation Engineering*, Vol. 1, pp. 54–6.

Toledo, P.E.C. de and de Freitas, M.H. (1993) Laboratory testing and the parameters controlling the shear strength of filled rock joints. *Géotechnique*, **43**(1), 1–19.

Turk, N., Grieg, M.J., Dearman, W.R. and Amin, F.F. (1987) Characterization of rough joint surfaces by fractal dimension. *Proceedings 28th U.S. Symposium on Rock Mechanics*, Tucson, AZ, pp. 1223–36.

Vijayvergia, V.N. (1977) Friction capacity of driven piles in clay, in *OTC Conference*, Houston, Paper No. 2939.

Wesley, L.D. (1975) Influence of stress-path and anisotropy on the behaviour of a soft alluvial clay. PhD dissertation, University of London.

Wroth, C.P. (1975) In-situ measurement of initial stresses and deformation characteristics, in *Proceedings ASCE Special Conference on In-Situ Measurement of Soil Properties*, Raleigh, NC, American Society of Civil Engineers.

Additional reading

Irving, C.D. (1978) *A Programmed Introduction to Principal Stresses and the Mohr Stress Circle*, The Institute of Materials, London.

Index